• HALSGROVE DISCOVER SERIES ➤

DEVON'S NON-METAL MINES

For Jenny and Katie

DEVON'S NON-METAL MINES

Discovering Devon's Slate, Culm, Whetstone, Beer Stone, Ball Clay and Lignite Mines

Richard A Edwards

HALSGROVE

First published in Great Britain in 2011

Text and images copyright © Dr Richard A Edwards 2011, except for images otherwise credited
in the captions, which are copyright of the organisations and individuals named.

The right of Dr Richard A Edwards to be identified as the author of this work has been asserted
by him in accordance with the Copyright, Designs and Patents Act 1988.

Any views or opinions expressed in this publication are solely those of the author and do not necessarily
represent those of the publisher.

In the interests of your personal safety and enjoyment, the publisher and author recommend that you
follow fully all the relevant Safety, Fossil and Countryside codes. Most of the sites mentioned in this
book are on private land. Do not enter without the permission of the landowner. The author and
publisher can accept no liability whatsoever.

All rights reserved. No part of this publication may be reproduced,
stored in a retrieval system, or transmitted in any form or by any
means without the prior permission of the copyright holder.

British Library Cataloguing-in-Publication Data
A CIP record for this title is available from the British Library

ISBN 978 0 85704 118 0

HALSGROVE
Halsgrove House, Ryelands Business Park,
Bagley Road, Wellington, Somerset TA21 9PZ
Tel: 01823 653777   Fax: 01823 216796
email: sales@halsgrove.com

Part of the Halsgrove group of companies.
Information on all Halsgrove titles is available at: www.halsgrove.com

Printed in Italy by Grafiche Flaminia

CONTENTS

ACKNOWLEDGEMENTS

A book like this is by its nature largely a compilation of the work of others who have gone before and recorded their thoughts and findings in many places. Consequently I have drawn on a wide range of sources (page 153), and the generosity of the many people and organisations that have provided me with information has been a source of great pleasure. I have tried to list all who have helped me, but I apologise if I have omitted anyone.

General. Staff at the Westcountry Studies Library in Exeter, the Exeter Reference Library, Exeter University Library, and the Record Offices in Exeter and in Barnstaple have been most helpful.

Chapter 2. Penn Recca. I am most grateful to Dr Trevor Shaw for permission to draw freely on his 1952 article in *Cave Science*. He has taken an active interest in my account, and has provided all the splendid photographs of underground scenes in the mine. I am grateful to him for reading and commenting on a draft of this chapter. David Checkley, Chairman of the British Cave Research Association, kindly gave permission to use copyright material in the article by Trevor Shaw. The Revd Nicholas Pearkes, Vicar and Rector of St Peter and St Paul, Broadhempston, has been most helpful in providing me with documents, illustrations and other information relating to Penn Recca. Others who have helped me include the Merchants of Higher Penn Farm, Janet Nash of the British Caving Library, David Wills of the Devon Bat Group, Peter Chapman of the Vincent Wildlife Trust, Roger Taylor, Jason Pain (Devon Spelaeological Society), Craig Dixon (Devon County Council Biodiversity Officer), Denise Ramsay and Amanda Newsome (Natural England), Jain Wood (Staverton History website) and Peter Glanvill.

Chapter 3. Culm. I am indebted to the Revd Richard Acworth for permission to use and quote freely from his article in the *Journal of the Trevithick Society* for 1991, and for providing additional documents. Members of the Trevithick Society, including the Curator, Pete Joseph, have been of great assistance. The Chairman Philip Hosken and Editor Owen Baker kindly gave permission on behalf of the Society to use the copyright material in Richard Acworth's article. Barry Hughes of the North Devon Maritime Museum has been most helpful, and has provided invaluable information, including rarely seen images. Pat Wiggett kindly arranged for the scanning of images. Felton Vowler and Clive Fairchild of the Bideford Railway Heritage Centre (BRHC) have assisted me greatly in providing photographs and other information, including scans from the *Atlantic Coast Express Magazine* (the house magazine of the BRHC). Ray Webster, Head of Environmental Health, Housing, Planning and Public Protection at Torridge District Council (TDC), and Jon Charles of the Building Control Department at TDC, were most helpful in allowing me to use data in reports and maps held at Bideford. I am most grateful to Richard Gould of Frederick Sherrell Ltd., Engineering geologists and Geotechnical engineers of Tavistock, for permission to use data contained in their reports on various sites in the East-the-Water area of Bideford. Others who have helped me include Peter Claughton, the Burton Art Gallery and Museum (Warren Cullum, Collections/Exhibitions Officer), Pat Slade of the Bideford and District Community Archive, Rose Arno of Bideford Library, and the British Geological Survey.

Chapter 4. Whetstones. My sincere thanks go to Robin Stanes, author of the standard account of the whetstone industry, who has very kindly allowed me to make extensive use of his work throughout this chapter. Many thanks to Philippe Planel, Parishscapes Project, East Devon AONB, for invaluable assistance, and for reading and providing helpful comments on a draft of the chapter. I am grateful to Richard Tamplin for allowing me to draw on his excellent study of the whetstone mining industry, particularly his account of the buildings and settlements of Blackborough and Ponchydown. Ramues Gallois kindly read through and commented on a draft of the chapter. Others who have helped me include Derrick Rugg, Ray Northam of Bodmiscombe Farm, Margaret Lewis of Allhallows Museum (Honiton), Mark Woods of the British Geological Survey, Judith Elsdon, curator of the Tiverton Museum of Mid-Devon Life, Christine Dunford, and Peter Grainger.

Chapter 5. Beer Stone. I am indebted to John Scott, proprietor and driving force behind the Beer Quarry Caves venture, for allowing me to draw on freely and quote passages from the booklet by himself and the late Gladys Gray. He has also been most generous with providing images of the Caves, many of which are reproduced in this chapter. Philippe Planel, the East Devon AONB Parishscapes Officer, kindly allowed me to use information about Beer Stone documents consulted at Clinton Devon Estate archives and in the possession of John Scott, details of which are included in his report on the Beer Quarry Caves Project. He also read and provided helpful comments on a draft of the chapter. Ramues Gallois kindly read through and commented on a draft of the chapter and provided illustrations. Chris Wood, formerly of Camborne School of Mines, now with Merrett Survey Partnership, very kindly produced a simplified version of his digital survey of Beer Quarry Caves for use in this chapter. I thank also John Scott (Beer Quarry Caves), Andrew Wetherelt (Camborne School of Mines), Bill Horner (Deputy County Archaeologist), Philippe Planel and Peter Youngman (East Devon AONB) for permission to reproduce the survey. Others who have helped me include John Torrance, Clinton Devon Estates (Estate Director John Varley and Archivist Gerald Millington), British Geological Survey, Denise Ramsay and Amanda Newsome (Natural England), and Peter Glanvill.

Chapter 6. Ball Clay. I am indebted to the Ball Clay Heritage Society and its Chairman John Pike and Committee Member Tony Vincent for invaluable help with this chapter. The Society kindly agreed to allow me to draw freely on its account published in the book *The Ball Clays of Devon and Dorset*. John Pike was kind enough to allow me access to the Society's fascinating and important archive and helped me to choose photographs from the archive with which to illustrate this chapter. John Pike also read and made helpful comments on a draft of the chapter. Tony Vincent, former Chief Geologist of Watts Blake Bearne, also read a draft of the ball clay chapter and let me have very helpful comments. My friend and long-time Geological Survey colleague Ted Freshney has shared his memories of the ball clay industry of north Devon with me. He also is responsible for producing map **6.1**, and the geological maps of the Bovey Basin and the Petrockstowe Basin, and modified them to suit this chapter. Michael Messenger's 2007 book on *North Devon Clay* has been an invaluable source of information for the Petrockstowe Basin. I am grateful to him for helpful comments on a draft of Chapter 6 and for permission to reproduce a figure from his book. I am grateful to the artist Frederick Cox for permission to reproduce his painting of a typical shaft mine in the Bovey Basin. Thanks to Stephen Adams, editor of *Quarry Management* magazine, for permission to reproduce figures.

Chapter 7. Lignite. Thanks to staff at the park office, Berkeley Park Homes, New Park, Bovey Tracey, for permission to photograph the Blue Waters Lake on 13 August 2009 and use the photograph in this book. Staff at the House of Marbles, Bovey Tracey, kindly gave permission to photograph the kilns of the former Bovey Pottery, also on 13 August 2009, and to use the photograph in this book. Others who have helped me include Felicity Cole (Curator of Newton Abbot Town and GWR Museum), and Christine Faunch, Acting Head of Archives and Special Collections at Exeter University Library.

Images. The sources and ownership of images in this book are generally given in the captions. Where no information is given the image belongs to the author. In the case of images 7.10, 7.11, 7.12 and 7.13, I have not been able to trace ownership.

Chapter 1
INTRODUCTION

ABOUT THIS BOOK

Much has been written about the celebrated metal mines of Devon, southwest England, from which tin, copper, silver, lead, iron and other ores have been won, but in this book I focus attention on the less well known, but perhaps even more fascinating, non-metal mines of the county. There are a remarkable number and variety of these mines, dug to exploit a wide range of rather surprising materials. They include the **slate mine** at Penn Recca near Buckfastleigh in south Devon; the 'culm' mines of north Devon – the county's only 'coalfield', in miniature; the **whetstone mines** of the Blackdown Hills in east Devon; the **Beer Stone mines**, also in east Devon; the **ball clay mines** of south and north Devon; and the **lignite (brown coal) mines** of Bovey Tracey, south Devon. Most of these mines are now forgotten, and little evidence remains of their former importance. Information about them is tucked away in many scattered and sometimes difficult to find places, and until now nothing has been written about them in one accessible book.

The materials dug from these mines have been extraordinarily variable. Just to take a random and incomplete selection, they have yielded roofing slates, whetstones for sharpening scythes and other tools, building stones, clay for ceramics, lignite for fuel and chemicals, and culm for burning as a fuel and for the manufacture of paint. A type of culm has even been used in the manufacture of mascara!

The chapters of this book constitute a gentle journey of discovery through different parts of Devon, during which we will look at the mines and their history, the methods used by the miners, and something about the miners themselves and their lives. In human terms it is a journey spanning 2,000 years, from the time when the Romans began to tunnel into the hills near Beer for Beer Stone, to the closure of the last ball clay mine in 1999. However these 2,000 years of mining history, although long by human standards, seem insignificant when we look at the span of time represented by the deposits that were being dug from the ground. These range in age from the slates of Penn Recca Mine that are around 360 million years old, to the lignites of Bovey Tracey that formed about 30 million years ago. Thus we will take a parallel journey into deep geological time looking at the rocks that contain the valuable materials sought by the miners. I will show how these rocks came into being, and what the world was like when they formed many millions of years ago.

My interest in writing about these mines came about through my career with the Geological Survey, much of it spent as a field geologist making geological maps in southwest England. There, over the years, I gradually came to find out more about the non-metal mines of Devon and developed an interest in them. When I came to look for another writing project in retirement, having just completed a book on the geology of the Jurassic Coast, I went back to the idea which had been germinating in my mind for some time, and decided to embark on writing a book about these fascinating mines.

A particular problem with studying the remains of mining landscapes is that so much disappears so quickly after the mines have ceased work. Much of the evidence on the ground has long since gone, but there are still clues to be found in the landscape. The search has taken me out into the beautiful countryside of many parts of Devon, ranging from the cliffs of north Devon, to the slate hills of south Devon, to the Blackdown Hills of east Devon, into the clay basins of north and south Devon, and finally into the stone mines of Beer. The other side of the detection process is the hunt for clues that have been recorded in writing, or on maps or photographs. This is

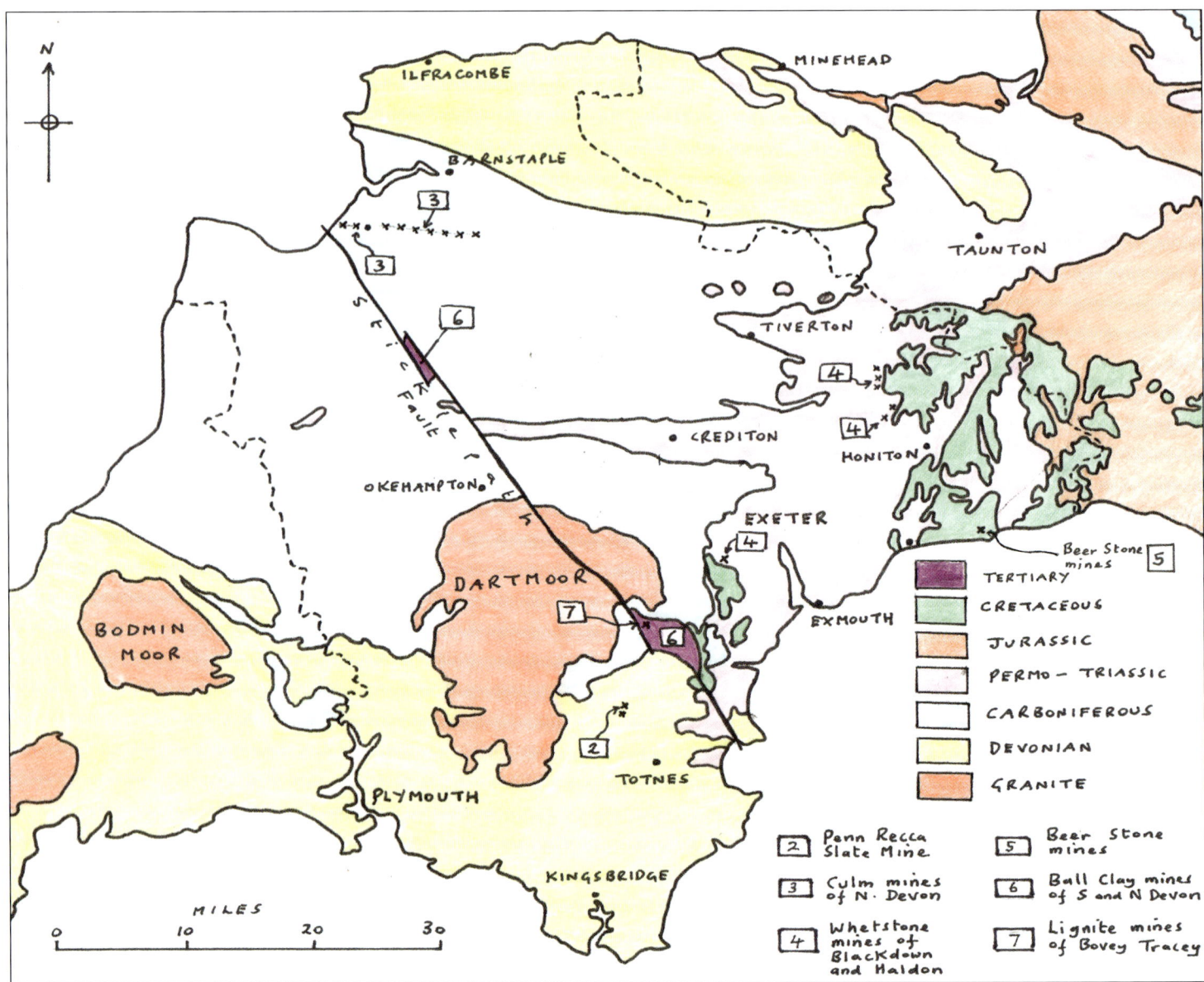

Legend:

- ■ TERTIARY
- CRETACEOUS
- JURASSIC
- PERMO – TRIASSIC
- CARBONIFEROUS
- DEVONIAN
- GRANITE

2 Penn Recca Slate Mine
3 Culm mines of N. Devon
4 Whetstone mines of Blackdown and Haldon
5 Beer Stone mines
6 Ball Clay mines of S and N Devon
7 Lignite mines of Bovey Tracey

1.1. This map of Devon shows the location of the mining industries explored in this book, and also the simplified geology of the county. The numbers in the boxes are those of the chapters dealing with each industry. The geological lines are based on the Ten Mile (1:625 000) Map, South Sheet, Third Edition (Solid) 1979, by permission of the British Geological Survey.

where the sleuth disappears into the library or the record room, or uses the resources of the internet.

I hope the book will appeal to a wide variety of readers. I have written it especially for the general reader, who may have an interest in one or other aspects of mines and mining, industrial archaeology, geology, history, or landscape, or who just likes a good story about some fascinating episodes in the development of the industrial history of the county. The book is easy to read and not too technical and I have included as many photographs, maps and other illustrations as possible to add to its appeal. This is not an academic work and is intended to entertain and inform in a pleasurable way – certainly no special geological or other knowledge is needed to read it.

INTRODUCING THE MINES

Since the earliest period of human history, people have tunnelled beneath the ground in search of an extraordinary variety of useful materials. In Britain, for example, miners were at work as long ago as 5,000 years before the present, up to 45ft beneath the surface, hacking out prized flints from the Chalk with picks made of red deer antlers. The surface of the Breckland landscape in Norfolk is pock-marked by mounds marking the tops of shafts dug for flints by these New Stone Age miners. I recall as a young boy – when there was a much more relaxed attitude to health and safety! – crawling to the very ends of these tunnels at Grimes Graves and wondering at the dark circles on the roofs, made by soot from fat-burning lamps, so many thousands of years ago.

Coming closer to home, southwest England is rich in valuable minerals, particularly those from which metals can be refined. The surviving remains of the famous tin, copper, silver, lead and other mines of Cornwall and Devon are so important that in 2006 the Cornwall and West Devon Mining Landscape was designated a World Heritage Site. Also well known are the silver and lead mines found in the north of Devon around Combe Martin.

But our book will focus on the less well known non-metal mines of Devon. Below is a brief preview of the places that we will visit, with some of the highlights of the trip. We will explore each mining industry in the order of age of the deposits that were mined, starting with the oldest. The locations of the industries in Devon are shown on the map, 1.1, which also shows the simplified geology of the county.

• *Penn Recca – Devon's only slate mine*. We begin in Chapter 2 of the book near Buckfastleigh in south Devon with a visit to Penn Recca, probably the only slate mine in Devon, mostly forgotten, where 360-million year old slates have been wrested from the ground since the 14th century.

• *Coal and paint: the culm mines of north Devon*. In Chapter 3, we move on geographically to the Bideford area of north Devon, and forward in time to about 320 million years ago. We usually associate coal mining with the great coalfields of Britain, such as that in South Wales just across the Bristol Channel from Devon, but it is perhaps surprising to learn that Devon has its own little 'coalfield' which lies in north Devon to the east and west of Bideford. Seams of material rich in carbon, locally called 'culm', have been worked from the area for hundreds of years. Some seams consisting of a type of coal called anthracite were dug for burning as a fuel, but another seam was mined, especially in later years, for use as a paint pigment called 'Bideford Black'.

• *The 'Scythestone Hills': the whetstone mines of the Blackdown Hills*. Next, in Chapter 4, we travel about 40 miles to the east and forward in time about 220 million years to the charming Blackdown Hills, which straddle the Devon-Somerset border south of Taunton. Over 100 years ago, the village of Blackborough on the western slopes of the hills was the centre of a fascinating and little-known industry where miners burrowed into the steep hillsides to bring out masses of sandstone concretions which were fashioned into sharpening stones, or whetstones.

• *'Cathedrals of stone': the Beer Stone mines*. In Chapter 5, we move in our journey of discovery from Blackborough to Beer in southeast Devon, where we find the fascinating Beer Quarry Caves. In this area, a special and unusual development of the Chalk, formed about 90 million years ago on the bed of a tropical sea, has been dug for building stone from subterranean caverns for about 2,000 years, from the time of the Romans.

• *Ball clay mines of north and south Devon*. We now move another 50 million years to a quite different and more recent period of earth history, about 40 million years ago. In Devon, layers of clay, sand and lignite were laid down in south Devon between Bovey

Tracey and Newton Abbot, and in north Devon around Petrockstowe. The clays include valuable seams, known as 'ball clay', which are widely used in the ceramic industry. We shall explore these deposits in Chapter 6 and concentrate on the workings that have exploited the ball clays from early times until the end of underground mining. This industry is the most economically important of all those that we will look at in the book and the last mines closed only in 1999, although open-cast working continues on a large scale.

• *Lignite mines of Bovey Tracey – the 'Bovey Coal'.* To end our journey of discovery we remain in the Bovey Basin of south Devon where we find the next valuable commodity that has been won from underground mines, described in Chapter 7. This material is called lignite, which, as mentioned above, is a type of low grade or brown coal. It has been worked south of Bovey Tracey where it has been dug not only in a large open pit (the Bovey 'Coal Pit' or 'Blue Waters Mine') but has also been mined from underground tunnels. The Bovey lignite was dug mostly to burn as a substitute for coal, but has also been considered as a source of organic chemicals, such as montan wax.

DIVIDING UP DEEP TIME

To appreciate the age and context of the deposits that have been worked in the various mining industries, we will look briefly at the way that geologists have divided up deep geological time. To begin with, we need to appreciate – although it is difficult to grasp the spans of time involved – that the earth is incredibly ancient - possibly over 4.54 thousand million (billion) years old. There is a great division in the history of the earth, which happened 542 million years ago. Then, there was a remarkable 'explosion' in the number of easily preserved shelly animals, now found as fossils in the rocks. Rocks older than that date, which represent the vast bulk (nearly 90%) of earth history, are called Precambrian, and contain few fossils.

Geological time has been divided up by geologists into parcels of time with special names. The last 542 million years of this geological time scale, with the named periods of time, and their ages, is shown in the table (1.2). The different periods are sometimes called after places where the rocks were first studied or can best be seen (a well-known example is the Jurassic period which is named after the Jura Mountains in France). The Devonian period is closer to home, and was first named in 1839 after the county of Devon by the famous geologists Adam Sedgwick and Roderick Murchison. Some of the periods are named after special characters of the rocks (for example, the Carboniferous is named from the fact that the rocks contain coal [*carbon*] seams). The Cretaceous is named after the Latin word for chalk ('*creta*') because of the British Chalk beds of this age. The table also gives some information about major events in the history of the world, and shows when certain important animal and plant groups first appeared on earth.

EARTH, THE EVER-CHANGING PLANET

It is interesting to think about the world we live in now, and how it relates to the world of the past. In order to understand the place that our various mining industries have in the geological story of the world we need to look at the way in which geologists have begun to understand the workings of planet earth. By the middle of the 20th century, a great deal was known about the geology of the earth, but there was no unified understanding of processes, and many features remained unexplained. All this changed in the 1960s, when there was a revolution in geological understanding. Studies of the ocean basins and the distribution of earthquakes led to the idea of 'plate tectonics' which recognised that the surface of the earth was made up of gigantic rigid 'plates' in constant, almost imperceptible, motion relative to each other, probably driven by convection currents deep in the earth. These movements explain the formation of mountains and the distribution of earthquakes, amongst many other things. The plates 'grow' at mid-ocean ridges and mountains form where two rigid plates collide. Geologists at last began to understand something about how the earth works.

At times the landmasses of the world have come together, while at other times they have been more fragmented. For example, about 240 million years ago, the world's landmasses were joined together into a giant 'supercontinent' called 'Pangaea' (meaning 'all lands'). About 200 million years ago, as a consequence of the movement of the great plates described above, Pangaea began to break up until, by about 100 million years ago (in the Cretaceous period – see table 1.2), the arrangement of the continents on the face of the globe was beginning to approach a form close to that of today, although there were still many differences. For example, a large ocean lay between Europe and Africa, and Europe and North America were much

Era	Period	Mines of Devon explored in this book	Age (millions of years)	The first appearance of:	Major events
Cainozoic	Quaternary (Holocene & Pleistocene)		1.8?	Man	Ice Ages in the northern hemisphere
	Neogene (Pliocene & Miocene)		23.0	Apes	Alpine mountains
	Palaeogene (Oligocene, Eocene & Palaeocene)	**Lignite mines of Bovey Tracey, Chapter 7** **Ball clay mines of the Bovey Basin and Petrockstowe Basin, south and north Devon, Chapter 6**	65.5	Monkeys Horses	Great extinction of life at end of the Cretaceous
Mesozoic	Cretaceous	**Beer Stone mines, east Devon, Chapter 5** **Whetstone mines of the Blackdown Hills, east Devon, Chapter 4**	145.5	Modern mammals. Flowering plants	Chalk widely deposited Dinosaurs on land
	Jurassic		199.6	Birds Dinosaurs	Ammonites abundant in the sea. Dinosaurs on land. Oceans widen.
	Triassic		251.0		Deserts. Great extinction of life at end of the Permian.
Palaeozoic	Permian		299.0	Theraspids	Deserts. Pangaea supercontinent
	Carboniferous	**Culm mines of north Devon, Chapter 3**	359.2		Glaciation. Variscan mountains. Coal forms in swamps.
	Devonian	**Penn Recca Slate Mine, Buckfastleigh, south Devon, Chapter 2**	416.0	Amphibians. Sharks	Fish and amphibians
	Silurian		443.7	Ammonoids. Jawed fish	Caledonian mountains
	Ordovician		488.3	Land plants	
	Cambrian		542.0	Trilobites	
Precambrian	Proterozoic		2500	Soft-bodied animals	Glaciations. Various periods of mountain building. Oldest glacial deposits.
	Archaean		≈4540		

1.2. The geological time scale. The ages of the boundaries between geological periods have been taken from The Concise Geological Time Scale *by Ogg and others, 2008.*

closer, since the sea floor spreading that gave rise to the Atlantic Ocean had only just begun.

A BRIEF LOOK AT DEVON'S EARTH HISTORY

We turn now to look briefly at the periods of Devon's geological history, including those that are relevant to our various mining industries. Looking at the simplified map of the geology of the county (1.1) we can see that it is largely underlain by rocks of Devonian and Carboniferous age, with younger rocks of Permian, Triassic and Cretaceous ages in the east (see table 1.2). The Devonian and Carboniferous rocks are arranged in a large downward bend or fold in the earth's crust, with the central part occupied by the Carboniferous rocks and the older Devonian rocks emerging to the north and south.

The Devonian. During the Devonian, which spans the period of earth history between 416 and 359 million years ago, most of Britain was land, lying between the equator and about 10° south latitude. However, in contrast, southwest England was the site of a shallow sea on the bed of which muds and limestones were deposited. The muds laid down on a late-Devonian sea-bed were the raw materials from which the slates of the Penn Recca Mine were later formed during the Variscan earth movements, as described below. This sea lay to the south of a major landmass – the Old Red Sandstone Continent – on which red gravels, sands and muds were laid down in an arid climate by rivers, and in lakes. The position of the shoreline was located generally along the line of the present Bristol Channel, but fluctuated in position, sometimes pushing farther south into what is now mid-Devon.

The Carboniferous. During the succeeding Carboniferous period (359 to 299 million years ago) Britain lay close to the equator, and the celebrated Carboniferous (or 'Mountain') Limestone formed in shallow tropical seas that extended over parts of Britain. Later in the Carboniferous, much of Britain was the site of sandy river deltas and coal swamps. Devon, however, lay beneath the sea for much of this period, with what is now the Bristol Channel still marking the position of the coastline, north of which was an area of coastal swamps and deltas. The unusual rock sequence in Devon, originally named the 'Culm Measures', began with limestones in the early Carboniferous and continued in the later Carboniferous with a thick

sequence of muds formed on the bed of a sea, into which were poured numerous pulses of turbid sand-laden material. The result was a distinctive formation made up of alternating layers of mudstone and sandstone which give a striped appearance to the outcrops in cliffs and quarries. In the north of Devon, however, there is a series of rocks which are more like the Coal Measures of South Wales, and which were laid down in cycles formed by river deltas repeatedly pushing out from the land to the north. Major coals did not form, but it is in this area that we encounter the 'culm' deposits which were mined near Bideford and which are the subject of Chapter 3.

The Variscan earth movements. Now followed a period of intense earth movements which are collectively called the 'Variscan Orogeny'. These movements affected the Devonian and Carboniferous rocks, and squeezed and contorted them into fantastic shapes ('folds') which are visible, for example, in the cliffs along the north Cornish coast. The Penn Recca slates that we will look at in Chapter 2 were formed by heat and pressure acting on the muds that were originally deposited on a sea bed towards the end of the Devonian, about 360 million years ago.

The Permian and Triassic. The highlands of southwest England which were pushed up during the Variscan earth movements began to be worn down actively in a desert climate at a time when Devon lay at latitudes of about 20 to 30° north of the equator. The resulting debris was washed down on fans, blown by sand, or laid down in lakes, to form the distinctive series of red rocks of Permian and Triassic age collectively known as the New Red Sandstone (299 to 200 million years old).

The Jurassic. The Triassic was followed by the Jurassic period (200 to 145 million years ago) when a shallow sea spread over much of the British Isles except for some upland areas, and widespread deposits of mudstone and limestone were formed. There are virtually no Jurassic rocks surviving in Devon, and the main occurrences can be found to the east of the county, along the celebrated World Heritage Site called the 'Jurassic Coast'.

The Cretaceous. Sea levels rose again in the Cretaceous, from about 100 million years ago, and a sea with a sandy bed covered much of southwest England, stepping over older rocks progressively farther westward. The resulting sands and sandstones are preserved today as the Upper Greensand Formation which we will look at in more detail in Chapter 4, for it is from this formation that the whetstones of the Blackdown Hills were mined. As sea levels continued to rise in the Late Cretaceous, the Chalk seas eventually drowned the whole of the land mass of the southwest peninsula. Within the Chalk at Beer, a unique locally developed layer called the Beer Stone has been worked since Roman times for building stone and we shall explore it in more detail in Chapter 5.

The Tertiary. The seas retreated with the beginning of the Tertiary period about 65 million years ago, and most of the southwest became land. On this land surface, basins formed along the lines of large northwest to southeast-trending cracks or 'faults', and were filled with considerable thicknesses of clays, sands and lignites laid down in lakes and by rivers. We will look at the commercially valuable ball clay seams and lignite occurring in these Tertiary rocks in Chapters 6 and 7.

The Quaternary. For completeness, we will look briefly at the final chapter in the geological story. The last two million years or so of earth history is a period known as the Quaternary, sometimes called The Ice Age. It was a time of tremendous climatic change, and temperatures oscillated between cold (glacial) periods and more temperate (interglacial) periods. During the cold periods in the northern hemisphere, great ice sheets and valley glaciers advanced from the north and then retreated. Between these periods of severe cold, the climate became at times even warmer than today. About 10,000 years ago, the latest ice sheet melted and the climate became more temperate.

SAFETY AND ACCESS

Before we start our journey of discovery through the mines – a brief word about safety and access. All disused mine workings are potentially dangerous. Do not enter any adits or shafts. Not all shafts and adits are recorded, and old shafts in woodland are especially hazardous. The Penn Recca Slate Mine and most of the old culm mine workings in the Bideford area are on private land. The site of an old culm adit (not always visible) on the cliffs at Greencliff, west of Bideford, can be visited using public footpaths. The Lower Combe area of Penn Recca is crossed by a bridleway over which there is a public right of way. Nothing much remains of the whetstone workings of the Blackdown Hills except for the extensive overgrown terraces of waste material thrown out from the tunnels. The sites of these remains can generally be explored from public footpaths, but some of the land with workings is privately owned. The Beer Quarry Caves are open to the public as a tourist attraction, and guided tours are available – a visit is highly recommended. There are no working mines left in the ball clays of the Bovey or Petrockstowe basins. Access for individuals to the open pits is not generally possible. Sibelco UK operates in south and north Devon; Imerys stopped working ball clay in the Petrockstowe Basin in 2004. All that is left of the lignite industry near Bovey Tracey are the water-filled workings of the Blue Waters Mine pit, now a lake (to which there is no public access) partly surrounded by housing. The old bottle kilns of the nearby Bovey Pottery can still be seen at the House of Marbles, Bovey Tracey, where there is also an interesting small pottery museum. The former course of the celebrated Haytor Granite Tramway, now part of the 'Templer Way', passes near the old lignite workings of Bovey Tracey and is well worth a visit.

Chapter 2

ECHOING CHAMBERS:
The Penn Recca Slate Mine, Buckfastleigh, South Devon

INTRODUCTION

We begin our journey of discovery through the little-known non-metal mines of Devon by travelling to an area about a mile east of Buckfastleigh in south Devon (see the map, 1.1, on page 10), where at Penn Recca we find the remains of what is probably the only slate mine in Devon. Here, quarrying for slate began as long ago as the 14th century. After an idle period in the early 19th century, the quarries were re-opened in 1845, and work began on a series of tunnels and large chambers from which slate was mined. It is these fascinating underground remains, where a total length of 5463ft of passages has been recorded, that are the subject of this chapter. However, before we explore the mine in more detail, we will look briefly at the slate industry elsewhere in southwest England.

Slate in southwest England

Many of us are familiar with the spectacular slate mines of north Wales, such as Llechwedd, and many others. Some of these contain enormous caverns from which vast quantities of some of the best roofing slates in the world were extracted, mainly in the Victorian era. In southwest England, there was nothing on such a scale, but slate was dug from numerous, generally small, open pits and larger quarries scattered over Devon, Cornwall and Somerset.

The history of 'blue slate' quarrying in south Devon, mainly from the Lower Devonian and Middle Devonian rocks south of a line joining Plymouth and Torquay, has been discussed by Anne Born. She wrote that by the late 12th century Devon slate was being extensively quarried for a variety of uses - mainly for roofs, but also for walls, paving, coping, steps, slabs, window sills, lintels and hearthstones etc. Slate production increased especially during the great periods of church building, and there was a thriving export market.

The celebrated landscape historian W G Hoskins, wrote (in his 1972 book on Devon) that probably the best-known slate quarries in south Devon are those at Charleton near the Kingsbridge estuary, which provided slate for the many late Norman churches of the area. Also noteworthy are the slate quarries at Buckland Tout Saints from which great quantities of slate were exported to Holland in the 18th century, and the large quarries at Mill Hill, Tavistock (still open today), from where slate was sent to France and the Channel Islands. Hoskins noted that the Devon slates tended to soften with weathering, and by the middle of the 19th century, many roofs had been replaced with the more durable Cornish slate from Delabole. Later in the 19th century, the development of the railway network meant that it was economical to import Welsh slate, which was, in Hoskins' forthright opinion '....a disagreeable and foreign material...'!. Competition from the Welsh quarries and mines led to the gradual closure of Devon's slate quarries, and by the beginning of the 20th century only one or two survived.

Compared to the great number of open quarries and pits, there are, as far as I can establish, only four places in the southwest counties where slate has been mined. These are: in Cornwall at Carnglaze and possibly in the cliffs between Trebarwith Strand and Tintagel; in Somerset possibly at Treborough; and in Devon at Penn

Recca. Before going on to explore the Devon mine in more detail, we will look at the others briefly. All the mines in the southwest worked slates which date from the Devonian period of earth history (416 to 359 million years ago), described in Chapter 1.

Carnglaze Slate Caverns near Liskeard, now open as a tourist attraction, is the only slate mine in the southwest to approach in scale the underground workings of north Wales. The slates worked there are of Middle Devonian age, hard, blue, and well cleaved. Also in Cornwall, there are reports of underground slate workings in the cliffs between Trebarwith Strand and Tintagel, and 'Hole Beach' is reputedly named after an adit in the cliffs which may lead to underground workings. We cannot leave Cornwall without mentioning the famous large open pit workings at Delabole near Camelford in north Cornwall where grey and greenish grey slates of Upper Devonian age are still worked.

Turning briefly to Somerset, there were quite extensive open slate workings at Treborough, within part of the Ilfracombe Slates of Upper Devonian age, but the main quarry has long been filled with rubbish and closed, although extensive tips of waste slate remain through which there is a nature trail. It is possible that there were extensive workings underground as well as those of the large main open quarry.

Penn Recca is probably the only true slate mine in Devon but, as Anne Born has noted, tunnels have been dug at several other quarries in Devon for two main purposes: firstly to provide drainage from the quarry, and secondly to make it easier to get access to the slate or to transport it from the quarry. At Buckland Tout Saints quarry an adit 200ft long was dug to drain water from the workings. Part of a tunnel can still be seen at Ludbrooke [SX 658 548], and may have been used for drainage but also for transporting slate out of the workings. A tunnel at Winslade Quarry [SX 780 422] was used for carting slate out of the quarry. There is a tunnel from the bottom of the quarry at Gurrington Quarry [SX 784 696]. At Beesands [SX 822 416] a short (10ft) tunnel links the main quarry with a small adjacent working. The small quarry at Flete Farm [SX 766 465] was linked to the bottom of the steep lane to the valley below by a short tunnel. Finally, Anne Born records that at Wood Farm near Ugborough [SX 665 557] a tunnel nearly 1000ft long links two parts of the quarry (now filled in).

The Penn Recca Slate Mine and quarries

The Penn Recca Slate Mine and quarries are situated in attractive hilly country, with far-reaching views in places, about one mile east of Buckfastleigh (photo 2.1). Here we find the fascinating remnants of a slate quarrying and mining industry which spanned a period of over 500 years, from the earliest known reference to the industry in 1388, to the closure of the mine in 1908.

The surface remains occur in two areas. To the north are the main

2.1. A general view of the country around Lower Combe in August 2009, looking south. The entrance to the Lower Combe adit (page 25) is situated towards the edge of the woodland in the small valley in the foreground.

2.2. This watercolour painting, possibly dating from the mid-1800s, shows the workings of Recca Quarry (although it is captioned 'Penn Quarry from the summit'). Image courtesy of the Revd Nicholas Pearkes.

quarries, long-disused, deep and overgrown, which are centred on grid reference [SX 763 673], about 440 yards north-north-west of Higher Penn Farm. To the south, in the Lower Combe area, about 440 yards southwest of Higher Penn Farm, is a wooded valley (see photo 2.1) from the northern end of which an adit [mouth at SX 7624 6670] extends northwards towards the main quarries. Details of both these areas are shown on Ordnance Survey maps dating from 1887, reproduced here as maps **2.5** and **2.6**. There has been some confusion in the naming of the various workings. For example, both these areas (the northern quarries and those around Lower Combe) are marked on the 1887 Ordnance Survey map as 'Penn Recca Quarry'. On the tithe map of 1845, the large quarries west of the lane 660 yards north of the Lower Combe adit are called 'Recca

Quarry', while the smaller workings east of the lane are called 'Penn Quarry'. Still more confusingly, however, in some documents the word 'Penn' or 'Pen' may be used for either quarry.

Beneath the fields that lie between these two areas are the underground workings of the Penn Recca Slate Mine. The workings include two large chambers from which much of the slate was mined, as well as a network of tunnels and adits which we will explore in more detail later in this chapter.

THE NORTHERN QUARRIES WEST OF THE LANE (FORMERLY RECCA QUARRY)

The quarries west of the lane (Recca Quarry) are the most extensive and consist of two deep excavations linked by a narrow cutting (see the map, **2.5**). A fascinating old watercolour painting (**2.2**) probably dating from shortly after 1847, shows the working quarry. We can contrast the imagined scene of noise and activity then with the silence and peace of the quarry today, with its near vertical walls covered with moss and the floors overgrown. There are overgrown tips to the south, and more open tips to the southwest which show grey slate waste with some quartz vein material.

The roofless remains of a building [SX 7640 6720] can be seen on the south side of the quarries. It has an internal width of about 11ft and an uncertain original length, possibly up to about 44ft, and there is a fireplace on one of the longer walls; its original purpose is uncertain. The 1887 map (**2.5**) shows the position of a chimney [SX 7640 6729] and other buildings immediately next to the lane. The chimney and other buildings were photographed in April 1949 (photo **2.3**).

2.4. J C R Reed at the Main Entrance to the Upper Series of tunnels in Recca Quarry on 13 April 1949. This entrance is now partly filled in and gated. Photo by Dr T R Shaw.

Opposite: *2.3. This photo, taken on 17 April 1949, shows the old chimney and nearby buildings of uncertain function, located beside the lane east of Recca Quarry.* Photo by Dr T R Shaw.

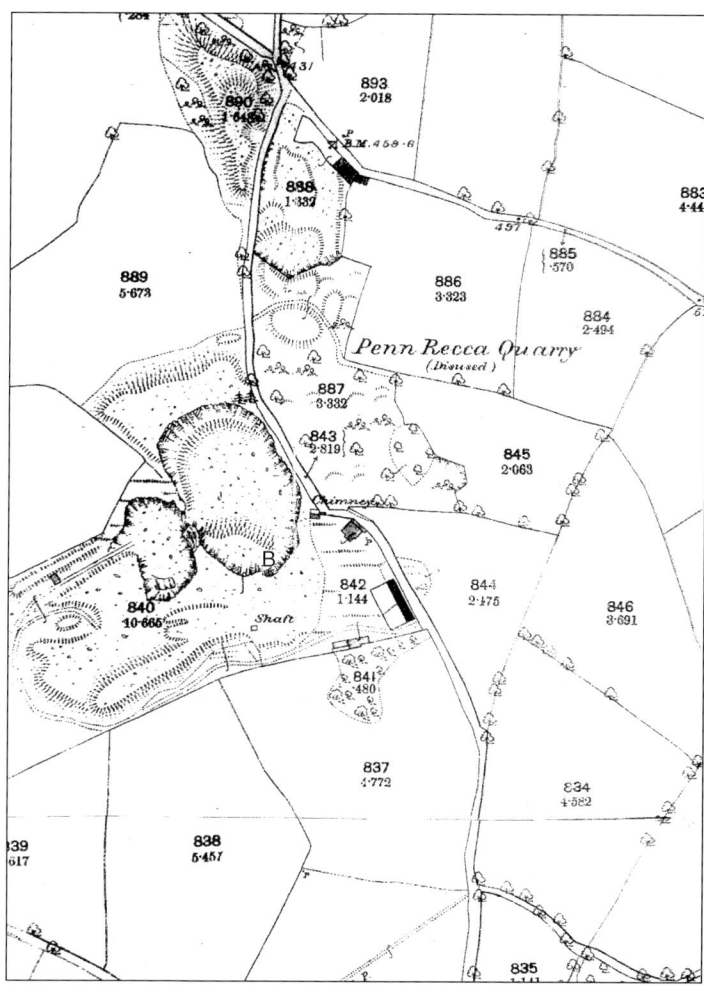

2.5. Part of a 1:2500-scale Ordnance Survey map of 1887 showing features of Recca Quarry (west of the lane) and Penn Quarry (east of the lane). Note the shaft south of the largest quarry and the chimney located next to the lane. The letter B indicates the position of the main upper entrance in Recca Quarry. Not to original scale.

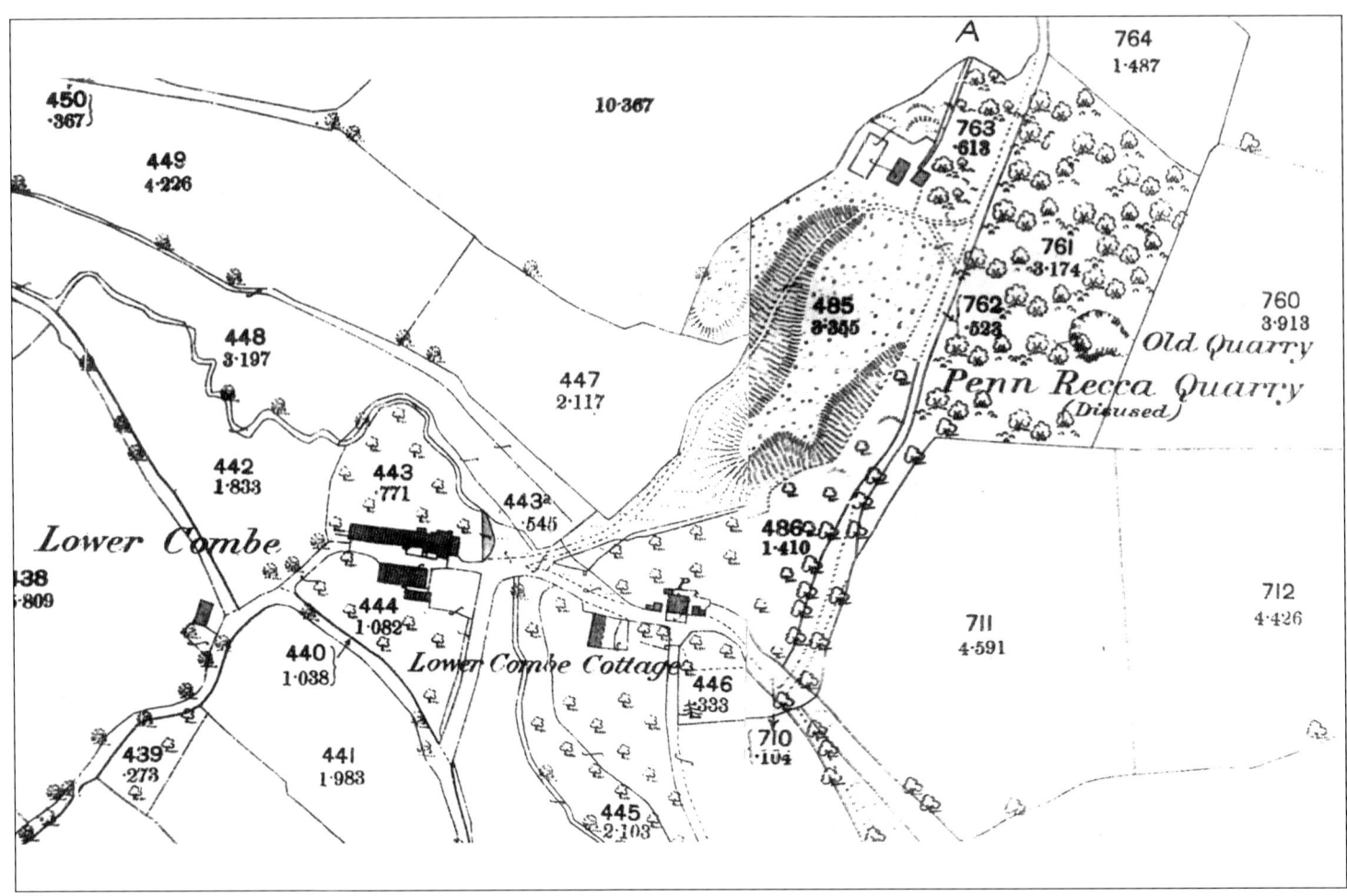

2.6. *Part of a 1:2500-scale Ordnance Survey map of 1887 showing features of the Lower Combe area. The mouth of the Lower Combe adit is marked at* **A** *on the map. Not to original scale.*

On the southeast face of the largest quarry (Recca Quarry) west of the lane is a gated entrance [SX 7638 6724] which is probably the Main Entrance to the Upper Series of underground passages (see below for more details). The photo (2.4) shows it as it was in April 1949. Another entrance, to the 'Fingerwrecker Passage' (described later), is present about 40ft south of the Main Entrance, but is now concealed beneath tipped material.

THE NORTHERN QUARRIES EAST OF THE LANE (FORMERLY PENN QUARRY)

These quarries, at one time called Penn Quarry, lie on the eastern side of the lane leading north from Higher Penn, and extend northwards to the cross-roads [SX 7636 6757] (see the map, 2.5). The main quarry, smaller than the workings of Recca Quarry, lies southeast of the cross-roads, and spoil heaps lie west of the cross-roads. In the southern face of the quarry there is an entrance to a system of tunnels which are unconnected to the main system.

Lower Combe

About 660 yards south of Recca Quarry is an overgrown area of old workings extending up a small valley from Lower Combe [SX 762 666] (see the map, 2.6). South of the entrance cutting are the remains of several old slate-built buildings (one is shown in photo 2.7), and south of there are tips of waste material (shown on the map, 2.6) dumped from the adit. The whole area is much overgrown. The waste material consists of roughly equal amounts of grey-green and purple slate, mostly small fragments, but with some larger pieces up to about 1ft across.

An old plan of the mid-1800s (2.8) is probably contemporary with the watercolour pictures shown in illustrations 2.2 and 2.12. It shows the line of the tunnel (the Combe adit) connecting Lower Combe and Recca Quarry, with the positions of shafts along it. In the Lower Combe area, the buildings marked are stables, office and carpenters' shop, and smiths' shop. In the Recca Quarry area a building is marked as an engine house and stable, with a smith's shop just south of it. Spoil heaps are shown in both the Lower Combe and Recca areas.

THE SLATES OF PENN RECCA

The Penn Recca area is underlain by a belt of purple, green and grey slates, trending roughly southwest-northeast, which are Upper

2.7. Part of a ruined mine building in the Lower Combe area, August 2009.

Devonian in age and belong to a formation called the Gurrington Slate. Thin bands of igneous rock – formed from the cooling and solidification of originally molten rock – occur within the slates of the long adit extending from Lower Combe to the large chambers of the mine, and there are also abundant quartz veins, particularly at the northern end of the workings. Fossils have been recorded from the Gurrington Slate, but they are generally very difficult to find, irregularly distributed and poorly preserved. They consist mainly of small bivalve shells and ostracods (a type of microscopic crustacean, sometimes known as 'seed-shrimps').

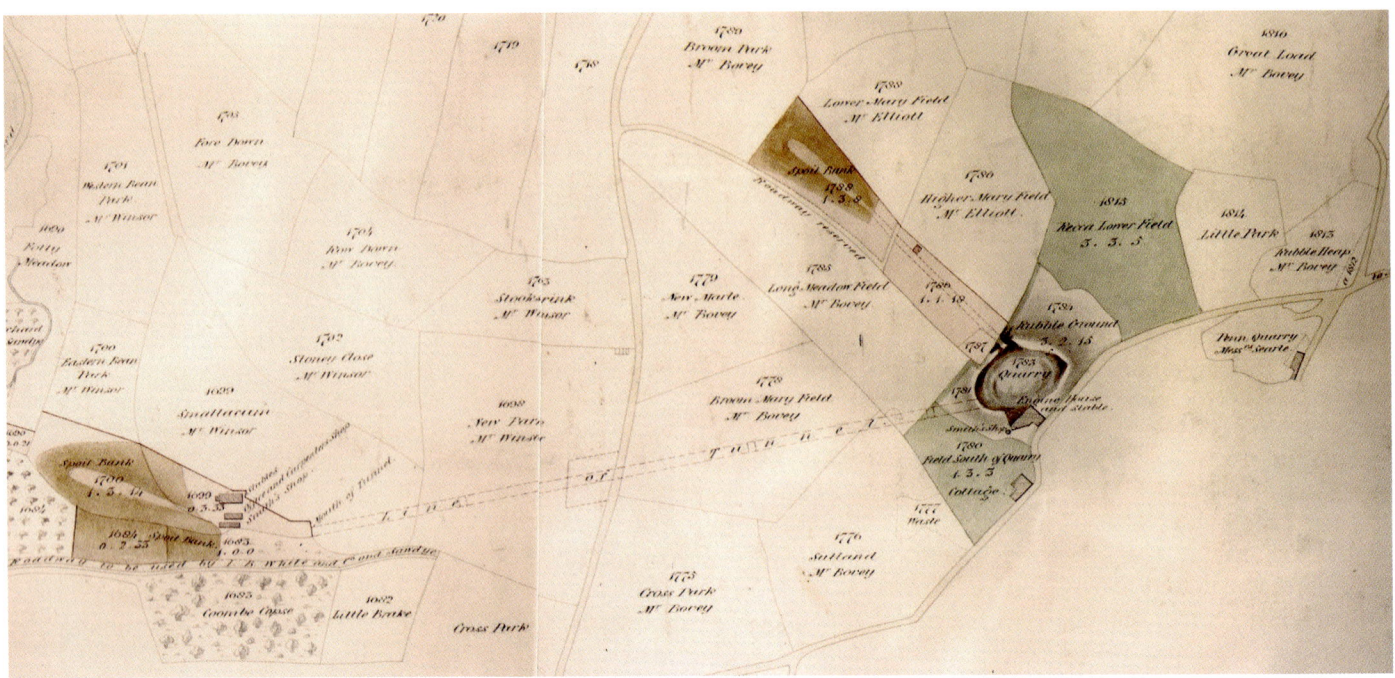

2.8. A mid-1800s plan showing workings in the Lower Combe area and the Recca area, and the 'Line of Tunnel' connecting the two. Image courtesy of the Revd Nicholas Pearkes.

The geology of the quarries was investigated in some detail by Dr Roger Taylor in 1995 as part of the scheme for recording RIGS (Regionally Important Geological and Geomorphological Sites). This site is a Devon County Geological Site, listed as RIGS site No 61. In the northernmost (Penn) quarries [SX 764675] Roger Taylor found glossy purple and green banded slate with the slaty cleavage dipping to the southeast at a low angle. In contrast, the large and deep (Recca) quarries on the west side of the lane (centred on [SX763 673]) are in uniform greyish-green slate in which the slaty cleavage dips at 16° to the southeast. The extensive tips on the southwest side of the quarry are of grey slate with some quartz veining.

The fields between the main quarries and Lower Combe show purple and green slate ploughed up in the soil, but the cutting near the mouth of the Lower Combe adit again shows greyish-green slate, in which the cleavage dips to the south-south-east at about 20°. Waste tips south of the adit mouth show purple and greyish-green slate, presumably brought out from the adit. Two small overgrown quarries on the east side of the Lower Combe valley also show uniform grey slate.

The origin of the Penn Recca slates goes back to a time between about 375 and 360 million years ago, during the Upper Devonian geological period. Then, as we have seen in Chapter 1, our area was a shallow sea, on the bed of which mud accumulated quietly, and hardened with time to become a rock called shale. By the end of the Carboniferous period, about 300 million years ago, movements of the plates making up the earth's crust crumpled up the rocks to produce great mountain chains in many parts of the world (for example, the Pyrenees, the Urals and, in North America, the Appalachians). In southwest England during these Variscan earth movements the Devonian and Carboniferous rocks were squeezed and intensely folded – we only have to go to the north Cornish coast (for example, Crackington Haven) to see places where the rocks have been buckled into fantastic shapes. The Devonian shales were buried and subject to increased pressure and temperature to change them from their original form – a process called 'metamorphism'. As the shales were squeezed and squashed, a series of changes took place. With increasing heat and pressure, various platy clay minerals within the shales became aligned at right angles to the direction of pressure, and also new minerals, such as micas and chlorite, formed. The new alignment of minerals formed planes of weakness, called slaty cleavage, along which the rock, now transformed into slate, splits or 'cleaves' readily. The cleavage is independent of the original sedimentary layers (bedding) that formed when the shales were first laid down, and slaty cleavage commonly cuts right across these original sedimentary layers.

EXPLORATION OF THE PENN RECCA SLATE MINE

Please note that entry to the underground workings of the Penn Recca Slate Mine is currently prohibited by the land owners (the Church Commissioners).

For about 40 years, from the date of the closure of the mine in 1908, until the explorations of 1949-51 (described below) there is no record of any exploration of the underground workings of Penn Recca, although doubtless they were visited at some time during this period. We owe almost all we know about the mine to the explorations of an intrepid group of cavers including W (Wilfrid) Joint (celebrated for his discovery with Mitchel and Northey of the Joint Mitnor cave near Buckfastleigh), T R (Trevor) Shaw, J C (Bob) Reed, E (Edgar) Reed, P (Pat) Cahill, S M (Murray) Gainswin, W M (Win) Hooper, and others. A detailed survey was made by Trevor Shaw and Pat Cahill in the course of a week in the spring of 1951, using a prismatic compass and metal tape, and a full account of the mine was published by Trevor Shaw in 1952. Trevor's account also gives more details about the progress of the 1949-51 explorations and those involved. More information is also provided in the *Devon Speleological Society Journal* for July 1989 where there is a fascinating account by E and J Reed entitled *Early days at Penn Recca – some excerpts from the HQ Log*' (for 29 and 30 January 1949 and 2 and 6 February 1949). These extracts record the progress of the exploration of Penn Recca, with many personal details which bring to life the

story of how the layout of the tunnels and chambers was established. For example, on 6 February 1949 the connection between the Upper and Lower series of passages was established, and is described as follows: 'From the top of the two shafts we were able to hear Squeak's [a curious nickname for Wilfrid Joint!] muffled voice a long way below, and by much shouting and stone dropping we eventually made contact with him, though not direct contact for a torch lowered down the 94ft shaft could not be seen at the bottom.......Edgar Reed and Squeak went to the Combe Quarry end of the Penn Recca workings to try and effect a joint (no pun intended) between the two (systems).'

In the years subsequent to the 1949-51 explorations, the mine was visited by numerous cavers and mine explorers, but few accounts have been left. One of the more significant events was the discovery on 14 January 1984 by Chris Proctor, Geoff Chudley and Chris Seddon of an extension to the known passages beyond the Main Chamber, which they called 'Stemple Passage' (see the map, **2.9**). The same passage was visited and described by eleven members of the Kent Underground Research Group on 27 September 2003. They found evidence of earlier visitors in the form of two inscriptions, one with the date '1971', and another with the name 'D Long 1979'.

In a letter dated 2 April 1974 to Trevor Shaw, Wilfrid Joint writes that the local press had reported that 'the chief instructor' of Outward Bound had been injured whilst exploring Penn Recca. Wilfrid went on to write that he hoped the caves would not continue to be used for such a purpose (that is, adventure training) which would damage the environment of the place, but should be scheduled as a protected industrial archaeology site to conserve the mine and its bat population. The site remains unscheduled, although of course bats now enjoy legal protection.

Another letter from Wilfrid Joint to Trevor Shaw, dated 13 November 1986, is worth quoting at length for the vivid picture it gives of a visit to the mine: 'Yesterday for the first time in over twenty years I went into Combe Tunnels. As you might know for some years masses of service personnel used the tunnels and shafts for abseiling practice but were stopped. To deter these and other folk the Combe entrance was walled up except [for] an over sized sheet metal door and the step heightened so that the water is 6" higher than the old high water level. You can I am sure envisage the scene yesterday at the entrance after all this rain there was a mini waterfall flowing down. The water was just an inch higher than my waders, so very reluctantly I had to discard them and do it the hard way. How I did enjoy it, after all these years, soon familiarised myself, but was disgusted at all the litter there... When I was in Combe I recalled the day we descended from Finger Wrecker cave that tin of flash powder you used [to take the photograph of Echo Chamber, 2.16]. Finger Wrecker entrance is some twenty feet under tons of mud tipped in the quarry and the same mass of mud has all but closed the main tunnel. The whole quarry is choked and overgrown with bramble trees etc. We are working there to see if the little entrance that is left is worth putting a grille on.'

In a letter to the magazine *Descent* of August/September 1992 the managers for the Church Commissioners for England (owners of the mine) expressed concern over the increasing numbers of groups making use of the caves, not all of which were led by suitably qualified personnel, and there was concern about possible pollution of the water supply which emerges from the Combe adit. Consequently, it was decided to block off both the entrances, at Lower Combe and the northern quarry, with grids which would still allow passage of bats. It was proposed that access to the caves would be restricted to one named caving society. The February/March 1993 issue of *Descent* magazine confirmed that the entrances to the mine had been gated and a key held by DCUC (Devon Caving and Underground Council) who would control access under certain conditions. In 2009, the Church Commissioners decided that access to the underground workings should be prohibited.

THE PENN RECCA SLATE MINE

Our detailed knowledge of the mine dates from the 'heroic' period of exploration between 1949 and 1951, described above. The results of the exploration and a detailed survey of the mine were given by Trevor Shaw in an invaluable account published in *Cave Science* in 1952. A further series of tunnels, totalling 413ft, beyond the Main Chamber were described in 1984 and in 2003. Much of what follows is based closely on Trevor Shaw's account, with his kind permission. Throughout, reference can be made to the plans of the mine reproduced here (2.9 and 2.10).

The tunnels are at two levels: the Upper Series is about 110ft above the Lower Series, and connected to it by two shafts and also through the roof of one of the large chambers. A striking feature of the mine layout is a long adit (the Combe adit of plan 2.9) which

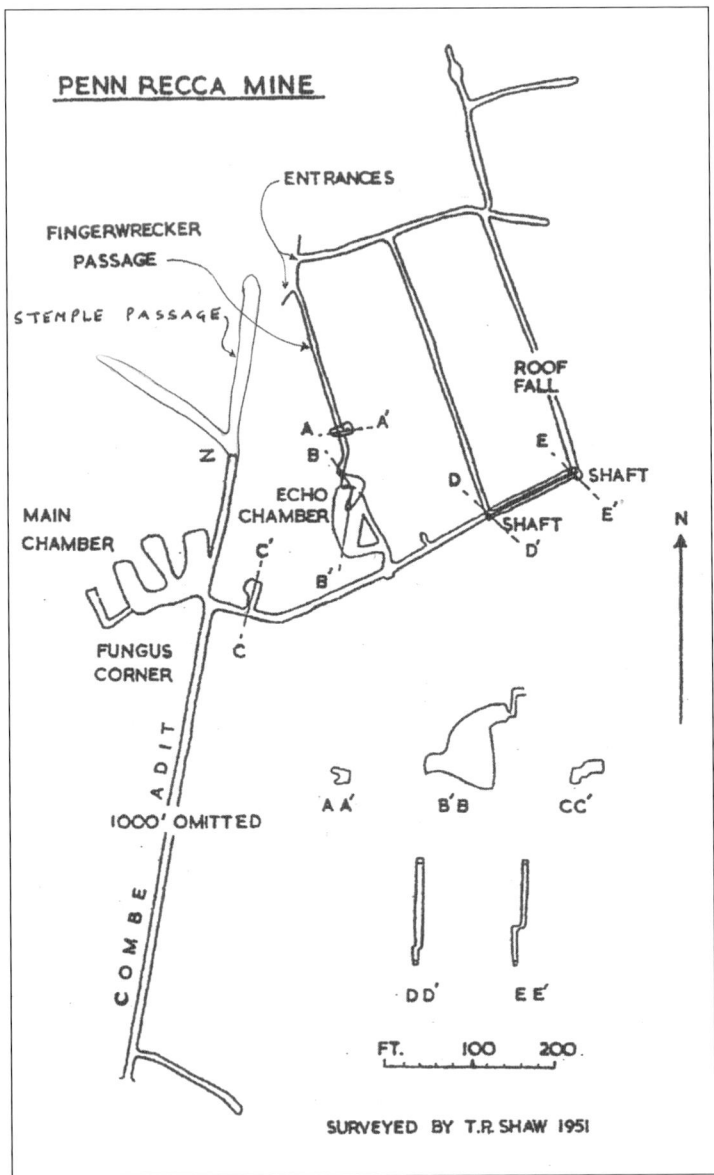

2.9. A plan of the Penn Recca Slate Mine, reproduced from a survey by T R Shaw and P Cahill in 1951, with additions after C J Proctor, 1984. Point Z indicates the beginning of the 'Stemple Passage' (page 27).

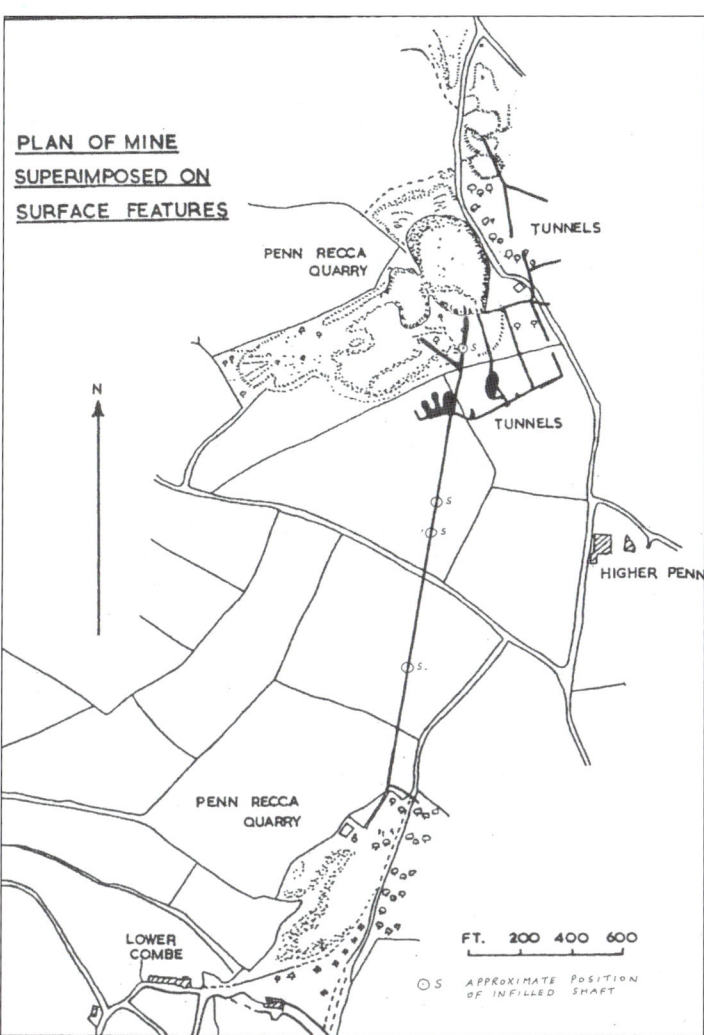

PLAN OF MINE SUPERIMPOSED ON SURFACE FEATURES

2.10. *A simplified plan of the Penn Recca Slate Mine, showing the relationship of the underground workings to surface features. Reproduced from a survey by T R Shaw and P Cahill in 1951. The approximate position of the 'Stemple Passage', based on a survey by C J Proctor in 1984, is also shown. The approximate positions of three shafts along the Combe adit, described by Dr T R Shaw (1952), together with another south of the main northern quarry, marked on an Ordnance Survey map of 1887 (see 2.5), have been added to Trevor Shaw's original plan.*

emerges in the southern (Lower Combe) area. There are also entrances in the northern quarries.

LOWER SERIES (COMBE TUNNELS)

Lower Combe adit

We begin with the adit at Lower Combe which was driven northwards into the hillside from the end of a small cutting in grey slate. The present appearance of the adit portal is shown in the photo (2.11) – the mouth is now covered by a metal plate secured by a locked metal bar. As in the days of the post-war explorations, the adit is still a source of water for properties in the area, and it is locked to protect the water source from pollution.

A fascinating watercolour picture, reproduced in illustration 2.12, shows the appearance of the area around the Combe adit mouth; it clearly post-dates the construction of the adit (1847) and

2.11. *The mouth of the Lower Combe adit in 2009.*

2.12. *A watercolour painting of the mid-1800s showing a scene at the mouth of the Lower Combe adit.* Image courtesy of the Revd Nicholas Pearkes.

2.13. *A view, taken on 18 August 1952, looking northwards along the Combe adit of Penn Recca Slate Mine.* Photo by Dr T R Shaw.

probably shows a scene shortly after that date. A tramway emerges from the adit and diverges into three lines – one is shown with a horse pulling a tub (or tubs) of slate. The buildings can be identified by reference to another plan (2.8), undated but probably contemporaneous with the watercolour of the adit mouth. The largest building on the left is a stable block. That in the centre is an office and carpenters' shop. On the right is the blacksmiths' shop.

The first 1550ft of the Combe adit is straight (see the plan, 2.10), and Trevor Shaw, writing in 1952, noted that even a quarter of a mile inside daylight could still be seen. The tunnel is about 6ft square throughout its length, and slopes gently down to the mouth. A mid-Victorian plan shows the gradient to be 1 in 142 from the adit mouth to a point 990ft in, and 1 in 176 for the remainder of its length. The photo (2.13), taken in 1952, shows a view looking north along the adit. About 40ft from the entrance, a tunnel branches off to the right, but it comes to a blind end after only 140ft. Its original purpose is uncertain.

Along the length of the adit, shafts were sunk from the surface. They occur at distances of 480ft, 990ft and 1108ft from the adit mouth, but there is no trace of them at the surface today. Another shaft, quite close to the main quarry, is shown on the 1887 1:2500-scale Ordnance Survey map (2.5), but cannot be located today. The

shafts along the adit were filled in by being bridged across the base with timbers and then rubble tipped in from above. Trevor Shaw notes that the bottoms of the shafts can be seen in the roof of the adit as timbers and boulders jammed together and cemented by stalagmite deposition. Below the position of each shaft there is a layer of stalagmite on the adit floor where lime-rich water has filtered down through the rubble of the shafts. Some of the original wooden tramway sleepers were seen in 1952, but were then fast decaying, and some had a coating of white stony material. It is uncertain when the shafts were filled in, but it was probably much later than when they were originally sunk. In September 1951 Trevor Shaw spoke to a 50 year-old man in Landscove churchyard, who thought that the infilling took place in about 1910.

The Main Chamber

At 1550ft from the adit mouth, there is a junction where the adit turns to the east at 90°. This junction, shown in photo 2.14, was named 'Fungus Corner' by Trevor Shaw after some massive growths of fungus in the stream that flows along the adit floor. On the left there is an archway through which we pass, over loose slate, into a corner of the Main Chamber, one of the main working areas from which slate was won (photo **2.15**). It consists of a wide passage about 80ft long and between 20 and 25ft high, from which three chambers of similar height and size branch off to the north (see the plan, **2.9**, which makes the layout clearer).

At the western end of the Main Chamber, there is a square opening in the wall, about 18ft above the floor. It leads to a passage about 6ft high which continues parallel to the lower wall of the Main Chamber for 39ft, then turns right up a slope for another 39ft. It seems likely that these passages were the preliminary stages in the formation of another working chamber which never materialised owing to the closure of the mine. Trevor Shaw noted that when this passage was first re-entered in 1951, they found the remains of a wooden ladder just inside. Other evocative finds were two barely recognisable candles fixed to the wall by daubs of yellow clay, the remains of a bucket, and a heart-shaped spade whose wooden handle had rotted away. Relics such as these bring to life the fact that real people with real lives toiled away hacking at the slate in these deep chambers.

2.14. 'Fungus Corner', Penn Recca Slate Mine, 18 August 1952. Photo by Dr T R Shaw.

2.15. A view of the Main Chamber, Penn Recca Slate Mine, photographed on 18 August 1952. Photo by Dr T R Shaw.

Stemple Passage

Another tunnel exits from the right-hand side of the Main Chamber when it is entered from the Combe adit. It is parallel to the Combe adit and almost in line with it. Trevor Shaw recorded that the end was reached after 120ft. However, later explorers have found an extension to the tunnels beyond this point. On 14 January 1984, Chris Proctor, Geoff Chudley and Chris Seddon followed the tunnel referred to by Trevor Shaw up a stream for 150ft where they found a collapse (point Z on plan 2.9), but were able to crawl through boulders to the right, 'dug and shored some years ago by persons unknown'. Beyond, they found a junction. On the left, a passage with a stream running along the floor came to a dead end after 200ft, with water pouring in from a small hole in the roof. This stream was at the time of their visit in 1984 providing most of the water flowing into the Combe adit.

However, straight on from the junction they found quite a different passage, fairly wide (in places up to 20ft, although normally about 15ft), low (stooping height), and strewn with slate boulders. A notable feature of the passage was the presence of a number of huge 'stemples' (wooden crossbars set between notches in rock walls, used as steps or for climbing) lying across it; niches in the walls showed where the stemples once spanned the roof, probably to provide working platforms. The passage continues for about 200ft where there are two massive collapses. The draught disappears into these collapses, suggesting a connection beyond to the surface, possibly up collapsed shafts. The ends of both the passages extend under or very near Recca Quarry, about 100ft above. The quarry was full of water before the development of the underground workings, so that the stream flowing down the side passage most likely drains the quarry above. I have added the position of the 'new' passages, based on the 1984 plan, to the original 1952 plans of Trevor Shaw (2.9 and 2.10). The 1984 explorers commented on the nature of this wide, boulder-strewn passage which is unlike others in Penn Recca; they believed that the slate boulders may have been carted in as waste from the Main Chamber.

The same passages were visited on 27 September 2003 by eleven members of the Kent Underground Research Group, and an account was given, together with a plan, by Harry Pearman. The explorers crawled past the same obstruction noted by the 1984 group of explorers (point Z on plan 2.9) which they attributed to a blocked shaft. The fork to the left was followed and the hole in the roof was attributed to a borehole. The passage straight ahead (the Stemple Passage of the 1984 explorers) was followed to the remains of a large quarried chamber ['Breakdown Chamber']. A large linked chain hung down from under a massive collapse choke. Two inscribed dates were found, reading '1971' and 'D Long 1979', indicating that these passages were known at least 13 years before the 1984 explorers recorded them.

Fungus Corner to Echo Chamber

The tunnel leading eastwards from Fungus Corner is again about 6ft square, and about 40ft along it on the left is found the partly-walled entrance to a chamber about 30ft long and 30ft high (see cross section C-C' on plan 2.9). Soon after the mouth to this chamber, the main passage turns left at about 37° and the stream flowing along the floor of the passage crosses to the opposite side; its source is not clearly defined, but lies between the corner and the next junction.

At this junction, the passage leads into the very impressive Echo Chamber (photo 2.16). There are two entrances: the one on the left reaches one end of the chamber over a steep slope; the passage below reaches the floor of the chamber at its own level after a few yards. Trevor Shaw describes the chamber as having an irregular pyramid shape, the base being about 80ft by 40ft, with a rounded top soaring up to 90ft above. A cross-section B-B' is shown on plan 2.9. Near the top of the chamber several wooden beams are wedged across, and there is a shelf just visible from below, which can be reached via the 'Fingerwrecker Passage' of the Upper Series of tunnels (described below).

The origin of the name 'Echo Chamber' is described by Trevor Shaw. In the summer of 1952, he and two others attempted to photograph the chamber using a large (3/16 lb) charge of magnesium flash-powder. He wrote that 'When this was ignited the bang was not unusually loud, but it set the air resonating throughout the whole chamber building up a pulsating and almost terrifying echo'.

Echo Chamber to the shafts

From the junction leading into Echo Chamber, the passage continues to the east-north-east for about 220ft. In 1952 it was blocked nearly up to the roof with piled slate, and there was another similar partial blockage 80ft farther on. The remaining 120ft of the passage lies almost immediately beneath a passage with similar orientation about 120ft vertically above, in the Upper Series, but the tunnels of the two series are linked by two shafts about 110ft apart. Both shafts have a curious kinked profile (see the cross-sections D-D' and E-E' on plan 2.9). The first shaft penetrates up for 16ft from the roof of the lower passage, then the centre-line is offset by about 5ft before the shaft extends up for another 94ft to the passage of the Upper Series; the second shaft (2.17) penetrates up for 40ft from the roof of the lower passage, then the centre-line is offset by about 10ft

2.17. Looking up the 40ft shaft at Penn Recca Slate Mine, 5 March 1949.
Photo by Dr T R Shaw.

Left: *2.16. The magnificent Echo Chamber, Penn Recca Slate Mine, photographed on 26 August 1952.* Photo by Dr T R Shaw.

before the shaft extends up for another 71ft to the passage in the Upper Series.

UPPER SERIES

The Upper Series of tunnels at Penn Recca Slate Mine, lying about 120ft above the lower, are entered from the southeast corner of the main quarry (Recca Quarry), about 700 yards north of the mouth of the Lower Combe adit (see plan **2.9**). The main entrance as it was in 1949 is pictured in photo **2.4**, but it is today almost blocked by debris. In 1952, it was 12ft high and 9ft wide, but soon reduced to a width a little narrower than the Lower Combe adit, and with a height between 5 and 7ft. The Upper Series of tunnels is a fairly simple network at right angles to each other (see plan **2.9**), and there are no large chambers as in the Lower Series.

It can be seen from the map (**2.10**) that the most northerly tunnel passes beneath the lane and ends close to one of the tunnels driven south from an entrance in the small quarry (Penn Quarry) on the east side of the lane, but there is no connection between the two sets of tunnels.

The mouths of the two shafts that lead to the Lower Series of tunnels can be seen at the southern end of the Upper Series. They are 6ft square and in 1952 were partly blocked by timbers. The shaft closest to the entrance (cross-section D-D' on plan **2.9**) goes down 94ft to a level floor, on one side of which is an archway which opens into the top of a similar shaft 16ft deep which connects to the Lower Series of tunnels (see above). The second shaft (cross-section E-E' on plan **2.9**) descends 71ft and at the bottom an opening only 18 inches square leads in to the top half of the lower part of the shaft, 40ft deep, which emerges into the Lower Series of tunnels (see above).

'Fingerwrecker Passage'

At the time of the 1949-51 explorations there was another opening in the main quarry (Recca Quarry), about 40ft south of the main entrance, but this is now completely blocked. In April 1949 a member of the exploration party injured his hand while clearing the entrance, and this incident gave rise to the spontaneous name of 'Fingerwrecker (Finga Recca) Passage' for the tunnel beyond! It was completely blocked at one time, and in 1949 there was only about 2ft clearance below the roof. Just inside there was a 4ft drop over slate rubble to a passage of 'normal' dimensions (that is, 5 to 6ft

square). About 240 ft from the entrance the tunnel ends and there is a square shaft which drops down 26ft to a level floor. A steep slope and another drop of 6ft, brings an explorer onto a ledge which opens out on one side to the vast space of the Echo Chamber, the floor of which is 72ft below. Above the shelf mentioned there is 18ft to the roof, giving a total height of 90ft for Echo Chamber.

TUNNELS EAST OF THE LANE (PENN QUARRY)

There is a separate series of tunnels in the area east of the lane (the Penn Quarry of the 1845 tithe map), the entrance to which is in the south face of the quarry [SX 764 675]. Trevor Shaw noted that the tunnels are level throughout, extending for 200ft at which point there is a junction with a branch tunnel on the left continuing for another 200ft. The main tunnel also continues for another 200ft (see plan **2.10**).

Trevor Shaw noted that there is nearly always a strong draught of air blowing through Penn Recca Mine. The air underground in winter is warmer than that outside, so that, because of the density difference, a draught is established from the Combe adit to the entrances above. In summer, the converse is true, the air underground is cooler, and the draught blows in the opposite direction. Trevor Shaw recorded a case where the air outside was very warm, and when entering the cave was cooled below its dew point so that a mist formed, and on 1 June 1951 fog formed in Echo Chamber.

THE HISTORY OF SLATE WORKING

Full details, including many passages quoted from original documents referring to ownership and leases for the various workings, are given in Trevor Shaw's 1952 account, to which the interested reader is referred. The earliest known reference to the slate quarries at Penn Recca was in a memorandum in the Chapter Act Book for 1388 when Sir John Holland, Lord of Dartington and a half-brother of King Richard II, was granted permission to roof 'the houses of the said manor of Dartington'. There is little evidence for the period intervening between the 14th century and the revival of the workings in the 19th century, but it is difficult to believe that in this very long period there was not some form of working, if only for local use.

The census of 1801 showed that the population of Staverton Parish was 1,053, comprising 473 males and 580 females, but by

1851, at a time when slate production reached its highest level, the population had risen to 1,152, made up of 562 males and 590 females. However, by the census of 1861 the population of Staverton parish had dropped to 949 which the census report attributed to the decline in employment in the slate industry, presumably because of competition from Welsh slate. The 1881 census report notes that there were still relatively large numbers employed in the slate industry, although agriculture remained the chief category of employment. The Staverton History website records that by 1845, when the mine was opened and expanded, over 400 people lived over two miles from Staverton Church. To serve these people it was decided to build a second church in the parish, near Thornecroft where most of the slate miners lived. At that time the land was a field called Landscore, changed to Landscove when the church was dedicated in 1851 in that name.

The tithe map of 1845 shows the quarries to be worked by two separate undertakings. The large quarries to the west of the lane were called 'Recca Quarry' while the smaller quarries to the east of the lane were called 'Penn Quarry'. The first separate reference to Penn Quarry is in 1823, at which date the quarry was rented to a Mr Searle (an alternative spelling for Searell). The quarry continued to be worked by the Searell family until 1852-3 when Richard Palk was the occupier, and he was still there in 1862. Trevor Shaw had no access to documents after that date, so could not trace the history of the quarry further. Between 1823 and 1845 over 9 million slates and over 170,000 rag slates were sold from Penn Quarry (a rag slate is a roofing slate with one edge untrimmed); the detailed figures are given in Trevor Shaw's 1952 account. Dues on slates were paid to the Dean and Chapter of Exeter at the rate of 6d for every thousand slates, and at 2 shillings per thousand feet of rag slate.

On the west side of the lane is the largest quarry, the Recca Quarry of the 1845 tithe map, although it is referred to by several names in various documents. As we have already seen, the entrances to the Upper Series of tunnels were driven from Recca Quarry. The quarry was under water for many years until 1845. A letter to the Dean and Chapter dated 5 January 1834, applying to work the quarry, notes that it had not been worked for the preceding 20 years. In 1842 a Mr Crace negotiated to work the quarry and was granted permission, but on 12 September 1844 it was noted that he had been 'overwhelmed with the expense thereof' and the assignment was passed to J B White. In a letter of 20 March 1845

White asked whether a lease could be granted '...before we determine on carrying out an expensive operation of driving a long adit into the Quarry, so as to work it after the Welch method'. On 14 July 1845 White was granted a lease of 60 years on the 'Penn or Recca Slate Quarry'.

In 1845 the Dean and Chapter of Exeter commissioned a survey of the mine by Richard Taylor of Pinner, near Falmouth; Eric Hemery, in his book *Historic Dart*, describes this professional approach as the planned prelude to expansion at the works. Eric Hemery notes another plan, also commissioned by the Dean and Chapel, which was prepared by Mr W Dawson of Exeter in 1850. It shows details of structures around the main quarry such as an engine house with nearby pump, stables, circular smiths' shop, and a saw pit. Tramroads to the tip are shown within the quarry. It seems that by 1847 the Combe adit was complete or nearly complete, for in the summer White was negotiating to build a road to carry slate from Combe.

A report in the *Mining Journal* for 1 December 1849 gave an interesting snapshot of developments at Penn Recca. The article was an enthusiastic piece praising the works and the qualities of the slates. It stated that nearly £30,000 had been spent in the previous seven years, a considerable part of which had been expended in tunnelling and open cutting. If correct, this is an enormous sum, equivalent to about £1,500,000 in today's money. The report continued: 'In our visit last week we were much delighted with the busy scene which presented itself, and more particularly with the ease with which blocks of many tons weight were quarried and removed to the end of the tunnel [the Combe adit], where they were all passed on to a weigh-bridge.... and thence conveyed to the slate works, where they were converted into roofing slates. After spending a very pleasant hour at this part of the works, we ascended the hill to the top of the quarry, where we found a number of men removing the surface rock, laying open fresh slate beds....'

By contrast, another article published in the *Mining Journal* only two weeks later (15 December 1849) was quite different in tone, and poured some cold water on the Penn Recca enterprise. The writer, Mr N Ennor of Treborough House, Washford, Somerset, claimed to have wide experience of quarry management in the west of England. He questioned the value of the large expenditure at Penn Recca, particularly the driving of the adit for half a mile at a cost of £15,000 (£750,000 in today's money). He wrote: 'I have not seen a body of good rock in the west of England sufficiently large to warrant so large an outlay, neither do I think it required'. Instead, he believed that use of a steam engine would have been far more efficient. He estimated that such an engine with all its gear, sufficiently powerful to raise 200 tons of rubbish and stone each day, could have been erected for £1,000 (£50,000 in today's money), Moreover, this would have taken three months to complete compared with the tunnel which in most slate rock would have taken three years to complete. He went on to say that 'experience has taught me that, under the present low prices of slate, there is no fortune to be made in the west of England quarries, without every branch of the work is carried out on the most economical principle, and with far less outlay than £30,000.'

The first few years of work by the new company were summarised in an entry in *White's Directory* for 1851 where it was confirmed that in the last eight years £30,000 had been spent in tunnelling and open cutting. About 100 men were recorded as being at work. It was noted that Penn Recca slate had been used for the roofs of farmhouses etc, 'since the time of Charles I and James I'. Eric Hemery noted, in his book on *Historic Dart*, that 'In 1870 the works were in the hands of a joint stock company who installed improved equipment and employed over 100 hands, by which time underground mining of the slate was well advanced and the great Coombe Tunnel had been driven. A fine chimney stack was built which still stands now partially ivy-covered'. In 1878 the quarry was still being worked, and *White's Directory* referred to an inhabitant of the village of Dean Prior being 'Captain of Pen Recca Slate Quarries, Staverton'. However, on Ordnance Survey maps of 1886, 1891 and 1906 the quarries are marked as disused. Possibly there was some intermittent working. Trevor Shaw noted that the last date of working at Penn Recca was 1908. A Penn Recca Quarry is mentioned in the Geological Survey Memoir of 1912 (*Geology of Dartmoor*) as showing greenish-grey smooth slate, but curiously there is no mention of underground workings, past or present.

In the 19th century, the main road from Ashburton to Totnes ran through Five Lanes, through High Beara to Bumpston Cross, which meant that the road was only about 600 yards from the Lower Combe adit. Thus, slate could readily be hauled from the mine to the high road and thence on the Ashburton or Totnes and beyond.

HOW THE SLATES WERE WORKED

Although it is generally more economical and safer to work slate from open quarries, in a situation where a valuable slate 'vein' is steeply dipping it may be necessary to work it underground. In either case, slate working was and is a dangerous business, but particularly so underground. Blasting was necessary, with the attendant risks – explosives and large pieces of rock below ground are a potent mix! The behaviour of slate in excavations is unpredictable, and masses detached during blasting were often likely to fall without warning. Dust was another problem, and this extended to the outside workings where dust was produced as a result of slate being cut or worked. We do not have any detailed information on how hazardous the Penn Recca workings were, but it would be surprising if there were not occasional injuries and possibly fatalities.

There are no first-hand accounts of the methods of slate working at Penn Recca, but they are unlikely to have differed from those employed in slate mines elsewhere. Large blocks of slate were won from the large underground chambers by drilling shot holes into the rock and blasting. The large slate slabs were placed on tubs drawn by horses along tramways with metal rails and pulled to the surface along the adits (see illustration **2.12**). In 1952 Trevor Shaw recorded, near the lower entrance to Echo Chamber, the only tramlines left in the mine, there being 20 inches between the inside edges of the rails.

Trevor Shaw has considered in his 1952 account the methods of working and the order of excavation of the mine. As noted above, the tunnels and chambers were excavated using blasting powder, for which shot holes were drilled to place the charges in the walls and roof. Many of these shot holes survive, and their orientation gives useful information about the directions in which the tunnels were driven. The tunnels of the Upper Series can only have been driven in from the surface, as confirmed by the direction of the shot holes. The upper parts of the two shafts that descend to the Lower Series of passages can also be seen to have been dug downwards. However, the lower parts of the shafts were worked upwards. It is a puzzle why the two shafts are misaligned, and why there are only small openings where they join.

Along the length of the Lower Combe adit, the direction of the shot holes changes several times. It appears that near the entrance to the adit, the tunnel was driven inwards, but farther in excavation proceeded outwards in both directions from the bottoms of the three shafts that extended to the surface.

It seems that Echo Chamber was worked from above, the workers reaching it along Fingerwrecker Passage. The shot holes were drilled downwards and the chamber increases gradually and steadily in size towards the bottom. In contrast, the Main Chamber has large areas of flat roof and must have been reached through the lowest point at Fungus Corner. Probably each arm of the Main Chamber was excavated separately in layers 6ft thick, the roof level probably first having been reached by blasting upwards at the entrance.

Trevor Shaw writes that the order in which the various parts of Penn Recca mine were excavated is largely conjectural, owing to the lack of historical references to specific parts of the mine. Construction of the Lower Combe adit was considered shortly after the larger part of the quarry was re-opened by J B White in 1845, and it was nearly finished by 1847. The sequence of excavation may have been as follows: firstly, the Fingerwrecker Passage was driven south from the main quarry (Recca Quarry) until good quality slate was discovered, and excavated in what eventually became Echo Chamber. The slate would have been taken out through the Fingerwrecker Passage and processed in the main quarry area. When the chamber was deep enough, and the quantity of good quality slate was realised, the Combe adit might have been driven so that the slate could have been transported from the mine more cheaply and efficiently, and possibly additionally to provide drainage for the workings. The Main Chamber was almost certainly excavated later, after completion of the adit. Slate brought out through the Combe adit would be processed in the Lower Combe area.

The date and purpose of the remainder of the Upper Series tunnels and the shafts connecting them with the lower tunnels are uncertain. Trevor Shaw speculated that if the upper tunnels and shafts were merely an elaborate way of looking for good slate, they may have been dug before the Echo Chamber, and the good slate there discovered from the bottom rather than the top as suggested above. Another possibility is that the Echo Chamber is late in date, and the adits and shafts were constructed just to move the slate from the main quarry towards the main Totnes Road near Combe, but this does not explain why the lower parts of the shafts were excavated upwards. It would have saved the cost of lifting slate out of the main quarry, but it seems an expensive option. The mystery of the two shafts and their misalignment is mentioned above.

One of the commonly known facts about slate working is that a great proportion (generally 90 to 95%) of the rock excavated ends up as waste. This explains why there are such large waste tips generally associated with slate workings.

HOW THE SLATES WERE USED

The property of splitting readily into thin layers, ease of working, and the fairly durable nature of the rock, makes many slates of great value. The better types of slate are especially suited for roofs, and during the Victorian era especially, vast quantities of Welsh slate in particular were used for this purpose. Poorer quality slate also has many uses, such as for building and for walling.

Slates come in many colours. The southwest England slates tend to be grey or greenish-grey. Delabole slate is commonly greenish-grey. The Welsh slates tend to be darker grey in colour. The Penn Recca slates are mainly greyish-green with lesser amounts of purple slate. An entry in *Kelly's Directory* for 1902, refers to distinctive 'sage-green' and 'chocolate' varieties of the slates – presumably referring to greyish green and purple slates respectively. In 1851, *White's Directory* records that 'The slate is found in immense blocks and is of a beautiful sage-green colour. Great quantities of this durable slate are now sent to various parts of the kingdom, and many of the farmhouses &c, in this neighbourhood have been roofed with it since the time of Charles I and James I. Ashburton Church was roofed in the former reign with slates from these quarries, and they remained till about 10 years ago'. Worthy repeated this information in 1875, and *Kelly's Directory of Devon* for 1902, under the entry for Staverton (page 644), again more or less repeats it, except that it is stated that the quarries are 'not at present used'.

WATER FROM THE SLATES

The water that flows from the Combe adit forms part of a supply for farms and other properties on the Staverton Estate. It feeds by gravity into spring catchpits. In 1952 Trevor Shaw referred to it as being the principal supply for much of Staverton Parish (an area of about 3000 acres). A letter to Trevor Shaw from the agents for the Church Commissioners, dated 30 October 1951, notes that for a much of the year there is sufficient to supply 30,000 gallons a day, but during the summer the supply had to be supplemented from other sources. During the winter there was frequently 'a rushing torrent' from the adit. However at other times the adit may be completely dry, as for example on 1 June 1951.

BATS IN THE MINE

Trevor Shaw noted that Penn Recca mine is a popular place for bats of several species. Over the period between January 1949 and July 1952 106 bats were ringed, as follows: Greater Horseshoe (*Rhinolophus ferrumequinum*) 84; Lesser Horseshoe (*Rhinolophus hipposideros*) 17; Natterer's (*Myotis nattereri*) 3; Daubenton's (*Myotis daubentonii*) 1; and Long-eared (*Plecotus auritus*) 1; and as well as these 145 bats were recorded that had been ringed elsewhere. The bats flew so frequently between various sites within 2½ miles distance, for example, the Buckfastleigh caves (nearly 1½ miles), Pridhamsleigh Cave (1 mile) and Bulkamore iron mine (2½ miles), that they could be considered as one living area for the bats.

In 1952 the bats seemed to prefer the Upper Series of tunnels, and only on a very few occasions had been seen asleep in the large chambers. In the Combe adit, bats were often seen hanging from the western wall, but never over the stream that flows on the eastern side. Bats had been seen to fly up the shafts connecting the Upper and Lower series of tunnels. I have made enquiries about the current state of knowledge about the Penn Recca bats. David Wills of the Devon Bat Group tells me (in September 2009) that, as far as he knows, there is no large colony of bats present in the mine tunnels, and that their main use is by small numbers of hibernating Greater and Lesser Horseshoe bats. The site has a high potential to be an autumn swarming site for a variety of bat species.

Chapter 3
COAL AND PAINT:
The Culm Mines of North Devon

INTRODUCTION

From the time of the formation of the slates that we looked at in the previous chapter, we move forward 40 million years and geographically from south to north Devon. Although we usually associate coal mining with the great coalfields of Britain, such as that in South Wales just across the Bristol Channel from Devon, it is a fascinating and little-known fact that Devon has its own small 'coalfield' which lies in the north of the county to the east and west of Bideford. Seams of material rich in carbon, locally called 'culm', have been worked from the area for hundreds of years, and certainly since the Middle Ages. Some seams of coal (a variety called 'anthracite') were dug for burning as a fuel, but another seam (the 'Paint Seam') was exploited, especially in later years, for use as a paint pigment called 'Bideford Black'. This intriguing and little-known small-scale industry, which we will explore in this chapter, ended in 1969 with the closure of the last mine at Chapel Park, east of Bideford.

The culm (coal and 'Paint') seams extend in a roughly east-west belt for about 12 miles from the coast at Greencliff, through the town of Bideford and East-the-Water, to at least as far east as Hawkridge Wood near Umberleigh, and are shown on the map, 3.1. The seams probably extend farther east into the parish of Chittlehampton. Other occurrences of culm have been reported from northwest and northeast of South Molton, south of East Anstey (on Oldways Moor), in the Hartland area, and near Cullompton.

DEVON IN THE LATE CARBONIFEROUS
320 MILLION YEARS AGO

To place the Bideford mining industry in context, we need to travel back in time about 320 million years, to the upper part of the geological period called the Carboniferous (page 14), when the land masses of the world were assembled into a giant supercontinent called Pangaea, surrounded by oceans. Vast coastal swamps extended across the equatorial regions of central Pangaea, covering parts of what are now North America, Europe, China and Russia. However, elsewhere on the globe, in great contrast, ice sheets covered much of the southern hemisphere; this Carboniferous 'Ice Age' lasted from about 330 to 260 million years ago. In the steaming tropical swamps, luxuriant vegetation thrived. Insects reached gigantic proportions: dragonflies as big as seagulls whizzed through the steamy air, and giant millipedes and cockroaches rummaged in the undergrowth. Amphibians and reptiles wallowed in the swamps. Giant clubmoss trees (lycopods) such as *Lepidodendron*, grew up to 130ft tall, and many different kinds of ferns thrived in the undergrowth beneath the trees. Especially important were tree ferns such as *Pecopteris*. Also common was *Calamites* which is related to modern horsetails and grew up to 70ft tall. Primitive types of seed plants such as *Neuropteris* and *Alethopteris* grew as shrubs or small trees with large fern-like leaves. Some examples of these plants are shown in illustration 3.2.

With time, the trees and other vegetation died and crashed into the swamps, building up to form great thicknesses of peat. The peats were altered by burial beneath sands and muds, involving increasing pressure and temperature, to form coal. It was the availability of this 'black gold' in the great coalfields of Britain that fuelled the Industrial Revolution. Depending on the degree of alteration, organic-rich layers range from soft peats, through lignite ('brown coal'), to sub-bituminous coal, through bituminous coal to anthracite. These changes result in an increasing proportion of

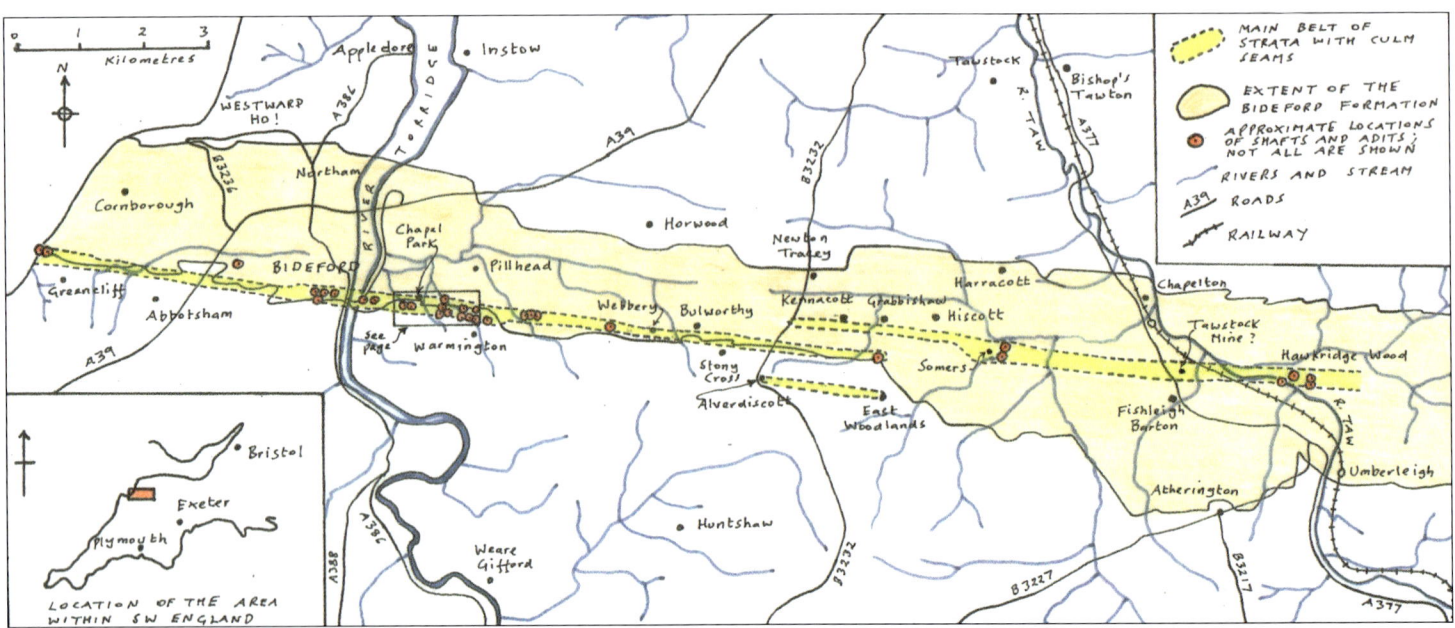

3.1. *This sketch map shows, in yellow, the main belts of strata with culm seams that extend from the coast at Greencliff in the west to Hawkridge Wood in the east. The seams are very variable in thickness and lateral extent and are not present everywhere along the coloured area. Not all shafts or adits are shown, and those marked are not precisely located. The positions of shafts and adits are based on information provided by Jon Charles and Ray Webster of Torridge District Council. The outcrop of the Bideford Formation, shown in brown, is based upon 1:50,000-scale geological sheets 292 (Bideford and Lundy Island) and 293 (Barnstaple),* by permission of the British Geological Survey.

3.2. *Some typical plant fossils from the Coal Measures (Upper Carboniferous).* Lepidodendron *(left) is a type of lycopod, a widespread and common type of Carboniferous tree;* Neuropteris *(middle) and* Sphenopteris *(right) are types of fern.*

carbon, so that peats have only about 65% carbon while anthracite has over 90% carbon.

The formation of coal takes place as part of a series of regular cycles. The coal forests were at intervals rapidly submerged beneath an invading sea (a process called a marine transgression), and thus sometimes the rocks directly overlying coal seams contain marine fossils. The shallow sea then silted up as river deltas pushed out from the land, and eventually the area became land on which another coal swamp formed. These cycles were repeated at least a hundred times in some places, and some geologists believe that the marine transgressions are related to rises in sea level during the periods between glaciations, when ice sheets melted causing sea level to rise all over the world.

The rocks around Bideford – ancient deltas and coal swamps

The rocks around Bideford form part of the great development of Carboniferous rocks that underlies much of central Devon (see map 1.1). The Carboniferous rocks of southwest England were named 'Culm Measures' in 1839 by the celebrated geological pioneers Adam Sedgwick and Roderick Murchison, the term apparently derived from the occurrence in them of sporadic thin beds of impure coal or 'culm' (the origin of the word is uncertain – it possibly derives from the Middle English word *colme*, probably equivalent to *col* (coal), plus the suffix *-m*, of uncertain meaning). The term culm is also used in South Wales (especially Pembrokeshire) to describe impure anthracites, and Walcot Gibson (in his book *Coal in Great Britain*) notes that 'In greatly disturbed and folded regions, as in Pembrokeshire and North Devon, where the seams of anthracite have been subjected to great pressure, the coal has been crushed to a fine powder or natural slack, which is called "culm"'. The Culm Supergroup is a modern name for the Culm Measures. It includes in its upper part two major subdivisions. The first and oldest is called the Crackington Formation after Crackington Haven on the north Cornish coast where the rocks, now crumpled by earth movements into spectacular folds, are beautifully displayed in the cliffs. The formation was laid down in a muddy-bottomed sea into which flowed at intervals sand-laden turbidity currents. The result is a distinctive layered sequence consisting of mudstone with many thin beds of sandstone. The second division of Carboniferous rocks that succeed the Crackington Formation is mainly made up of thicker sandstones, and is called the Bude Formation after the north Cornish coastal town.

The Carboniferous rocks around Bideford differ somewhat from both these types, and so have been given a separate name, the Bideford Formation. They are closer in many ways to the Coal Measures that are so well developed across the Bristol Channel in South Wales. They date from about 315 to 320 million years ago, and were laid down at roughly the same time as the mainly coal-free lower part of the Lower Coal Measures of the South Wales coalfield.

The main belt of culm seams that have been worked generally occurs near the local base of the Bude Formation which overlies the Bideford Formation. The most characteristic feature of the Bideford Formation is that the rocks are arranged into cycles, of which at least nine are present in the Bideford area. At the bottom of a typical cycle is a black mudstone which passes upwards into grey silty and sandy mudstones, then siltstones and sandstones, ending at the top of the cycle with a thick bed of sandstone. Such a sequence, in which the sediments gradually become coarser in grain size upwards, is characteristic of beds that have been laid down on a delta. We can imagine rivers flowing from the land to the north, in what is now Wales, and building out a series of deltas into the sea that lay to the south. At least nine separate deltas built out in turn to form the cycles that we see today.

The culm seams that are of particular interest to us are found at the base of the Bude Formation above a sandstone bed (named the Cornborough Sandstone) present at the top of the uppermost cycle of the Bideford Formation. However, very thin culm beds, in some cases only a few inches thick, may occur within the Bideford Formation. For example, Felton Vowler's 2004 map of the area between the coast and the River Torridge shows culm seams scattered through the Bideford Formation at a dozen or so levels. Map 3.1 shows the position of the main belt of culm seams in relation to the outcrop of the Bideford Formation.

ANTHRACITE AND 'BIDEFORD BLACK' SEAMS

Two types of culm seam have been worked in the Bideford area. The first is usually referred to as 'anthracite', a type of high-grade coal typified by very high (over 90%) contents of carbon. The second type is a mudstone or shale rich in carbon, and it is this seam, the 'Paint Seam' which when dried and ground-up produces the paint pigment known as 'Bideford Black'.

A feature of all the culm seams is that they vary greatly in thickness, locally over very short distances, from only a few inches to

10 feet or more. In the Chapel Park area east of Bideford, the productive seams vary between 13.6 and 0.4ft thick. This variability in thickness may be in part original, but is also due to thinning and thickening of the coals during the intense folding and squeezing that they have undergone since they were first laid down. The low-strength coals formed weak zones along which they were deformed during folding. The seams, which were originally laid down horizontally, have been tilted by the Variscan earth movements (page 14) and are now very steeply inclined. They were buried to tremendous depths of between 4 and 5 miles during these earth movements. The seams are displaced in places by faults (cracks in the rocks along which movement has taken place), mostly trending NNW-SSE, which are so common in this area. One glance at the Geological Survey map (Sheet 292) shows the numerous faults that affect the rocks of the Bideford area.

There are apparently only two main seams near the coast; the lower one consists of soft anthracite about 3.3ft above the Cornborough Sandstone at the top of the Bideford Formation, and the higher one, 330ft farther up in the succession, is probably the Paint Seam. Farther east, in the East-the-Water area, however, there appear to be four anthracite seams occurring north of the Paint Seam which is geographically the most southerly seam and forms the youngest of the productive seams in the succession. The apparent difference in the number of anthracite seams may be due to the faulting-out of three seams in the coastal area, or the splitting of a single seam eastwards to give the four seams in the East-the-Water area. However, because of the intense folding that has affected the rocks, producing steep and variable dips, it is not impossible that there are in fact fewer than four coal seams, repeated by folding.

Adam Sedgwick and Roderick Murchison noted in 1840 that about two miles east of Bideford (probably in the Chapel Park area), three highly inclined culm beds were worked, designated by the names of South, Middle and North veins. The *South Vein* or *Paint Vein* was about 3ft wide and on its north side was a carbonaceous shale used for paint. The whole bed was very impure and contained little good culm fit for burning. The *Middle Vein* occurred 90ft north of the South Vein and was about 4ft thick on average but expanded in some places to 20ft and in others pinched out to a few inches. The culm occurred precisely as in the contorted 'culmiferous' strata of Pembrokeshire, in large spherical masses, much shattered and broken. Some parts of this bed were reported to be very good for

domestic use. The *North Vein* was 240ft north of the Middle Vein, and was about 2ft wide on average, but liable to similar variations in thickness as the Middle Vein; it had been extensively worked though was not as high in quality as the Middle Vein. The footwall of all the veins was sandstone while the roof was generally shale.

In a report of 1927 by Humphrey Morgans (of which more later), the seams of the Chapel Park area were named as follows, in ascending order in the local succession:

The Paint Seam
(*South Vein* or *Paint Vein* of Sedgwick and Murchison).

No 2 Mary Ann Seam

No 1 Mary Ann Seam
(No 1 and 2 Mary Ann seams are closely spaced and may together be the equivalent of the *Middle Vein* of Sedgwick and Murchison; also called the *Great Anthracite Seam* by De la Beche in 1839).

'5-foot' Seam
(possibly the *North Vein* of Sedgwick and Murchison).

'2-foot' Seam.

The anthracite seams consist of coaly material, generally very much mixed up with black shale which wraps around the coal. The coal itself is generally in a shattered condition, even virtually a powder in some cases. Carbon contents from analyses of seams in No 1 Shaft in the Chapel Park area presented in the report by Humphrey Morgans are as follows: No 2 Mary Ann Seam: 80.00%; No 1 Mary Ann Seam: 83.53%; '5-foot' Seam: 56.82%; '2-foot' Seam: 67.56%. In No 2 Shaft, the '5-foot' Seam had a carbon content of 85.08%. The positions of shafts 1 and 2 are shown on the map (**3.10**). The Paint Seam consists of about one-third carbon, one-third alumina, and one-third silica.

Several writers recorded fossil plants in beds associated with the culm horizons. Henry De la Beche's 1834 discovery of plant fossils associated with the culm seams triggered off a major controversy which eventually led to the establishment of the Devonian system by Roderick Murchison and Adam Sedgwick in 1839. Townshend Hall, writing in 1875, gave a list of 26 plants, including the well known

Calamites (horsetail, an ancient genus which has survived to the present day), *Lepidodendron*, *Neuropteris*, *Pecopteris*, *Stigmaria* and others (some of these are shown in illustration **3.2**). In 1840 Adam Sedgwick and Roderick Murchison noted the presence of fossil plant remains in the roof shales at Chapel Park.

How the culm seams formed

Typical coal seams in most coalfields were formed in place by the decay of vegetation, and beneath a coal seam there is nearly always a fossil soil or 'seatearth' which contains rootlets extending down from the vegetation which grew in the coal swamps. Fragments of pale grey seatearth with rootlets have been found beneath a culm bed visible in the cliffs at Greencliff, indicating that vegetation grew in place. However, as far as I know, no other seatearths with rootlets have been found beneath the coals of the Bideford area. It is difficult to decide whether the coals of the area all formed in place, or if some consist of vegetation (tree trunks etc) which was drifted into its final location by streams flowing over a delta. The presence of the seatearth at Greencliff and the fact that the seams can be traced over a distance of 12 miles, does seem to suggest that they were formed in place, since it is difficult to envisage drifted material occurring consistently at the same horizons over such a wide area.

THE HISTORY OF CULM WORKING

Most of the mining activity was in the 19th and 20th centuries. The date of the first working of culm is uncertain, but during the Middle Ages anthracite for burning was probably dug on a small scale from surface outcrops or small pits. Alison Grant and Peter Christie, writing in *The Book of Bideford* (1987), show that there were active culm mines in the 17th century, for they note that the Bideford registers record 'A welchman a Collier' buried in 1629, and an entry for 1655 records 'John Boorges, Collier, found deade in the Culme worke'. Mining may have started in the area of Bideford High Street as early as the mid-17th century. The main mine in Bideford, in Pitt Lane, was 'waterlogged' by 1680, leading to the import of coal from Wales. The coal ships returned to Wales with cargos of items for the Welsh dairy industry such as butterpots, milk pans and pitchers.

A newspaper report of July 1793 refers to a mine at Tawstock. In 1807 the *Exeter Flying Post* of 25 November advertised a lease of a 'Culm Mine, lately discovered' at Westwood. When Chapple [Chapel] Park was put up for sale in 1818 (*Exeter Flying Post* of 28 May), the 'valuable Culm Mine' was referred to only after a description of the land and timber. These two advertisements suggest that the culm mining industry in this area was at this period emergent rather than long-established.

In 1808 Charles Vancouver described a bed of culm or anthracite in the parish of Chittlehampton, varying in thickness from 4 to 12 inches. He wrote: 'Some years since a vein of culm appearing near the surface on the parish of Chittlehampton, induced trials to be made of its quality, which proving satisfactory, a company was formed for the purpose of working it. This association continued for about two years, during which time five or six weys of culm were raised, and which was usefully employed in burning lime. [The definition of wey apparently varies with different types of goods; it might be two hundredweight (224lbs), or it could mean simply 'a wagon load']. The vein varied from about four inches to one foot in thickness; held a course correspondent with the general lay of the strata through the country, and dipped to the southward about one foot in three. The disappointment to several of the party, by the thinness of the seam, and the extraordinary expense consequently incurred in raising so small a quantity, occasioned the relinquishment of the project, although great expectations are still entertained in the neighbourhood, that the attempt will be renewed on a future day'.

The culm beds were shown on the geological map of England and Wales published in 1809 by Greenough. An attempt to show the distribution of the culm beds was also made on the 1809 edition of the one-inch Ordnance Survey map.

In the 1811 *Topographical Dictionary of England* by Samuel Lewis, the entry under Bideford notes that 'Mines of culm and black mineral paint are found in the vicinity, and on the rectorial glebe; the old culm mines have been lately re-opened, with every prospect of advantage'.

The Lysons, in their *Magna Britannia* of 1822, recorded anthracite on the coast near Bideford (presumably at Greencliff), and noted that it occurred as a near-vertical bed in the 'greywacke formation' and extended inland for many miles in a straight direction. The thickness was recorded at between 2 inches and 2 feet. Its use as a pigment at Plymouth Dockyard was noted. They wrote that, at Tawstock, [coal] was '.....procured in great quantities, and of a good quality, about the middle of the last century [1750s]. The works had been given up, and reopened about 1790. They were abandoned about 1800, on account of the water; at that time about

900 bushels a week were procured, the depth of the pit being then about 25 fathoms [150ft]'.

Henry de la Beche, writing in his celebrated *Report on the geology of Cornwall, Devon and west Somerset* of 1839, gave probably the fullest account of the anthracite or culm beds to date. He noted the extent of the beds for 12½ miles from Greencliff to Chittlehampton. He was obviously not too impressed, for he stated that the importance of these beds 'has been most remarkably magnified'! He noted that they were generally accompanied by black shales among some of which were found the abundant remains of plants. In 1840 the geological pioneers Adam Sedgwick and Roderick Murchison published an important description of the Carboniferous (Culm) strata of Devon, including quite a comprehensive description of the culm beds of the Bideford area.

In the East-the-Water to Chapel Park area culm mining began in the 18th century (possibly earlier), and the mines at Chapel Park and Westwood were established concerns by the late 1820s. Three seams were worked: the North Seam, Great Anthracite Bed or Middle Seam; and the Paint or South Seam. The Chapel Park Mine (and possibly the Westwood Mine) was taken over in 1846 by the Bideford Anthracite Company. The Chapel Park Mine closed in 1969.

White's *History, Gazetteer and Directory of Devonshire* of 1850 notes that 'Brown and grey paint and mineral black are got in the neighbourhood [of Bideford]; and at Chapple Park is the valuable culm mine of the Bideford Anthracite Mining Company, lately established, and now employing a considerable number of hands. A tram road, more than a mile in length, is being made underground at the heart of the mine.'

There seems to have been a decline in anthracite mining in north Devon from about the middle of the 19th century. The census returns (see below, page **53** and table **3.14**) record the reduction in the numbers of people involved in the industry. The reasons were probably various: competition from imported coal of higher quality; reduction in the demand for lime burning; exhaustion of the most easily worked seams; and problems with water at deeper levels in the mines. Probably also there was a drop in the demand for mineral black as wooden ships were phased out of use.

In the 1920s there was a revival in interest with the formation of the Devon Anthracite Company, the managing director of which was Mr Thomas Thornton and the manager Mr Rogers. They had ambitious plans to revive anthracite and mineral black mining and

took out leases along the whole outcrop of the culm seams. They sank a new shaft and opened old adits. A report was commissioned from the mining engineer Humphrey Morgans, who recommended that a colliery be established in Cleave Wood - more details are given on page **48**, below.

THE CULM MINES

As we have seen, the main series of culm seams extend between the coast at Greencliff, through Bideford, and then eastwards for about 9 miles to Hawkridge Wood near Umberleigh. The locations of the main mines are shown on the map (**3.1**). Very few surface remains can be seen today. There were many pits and quarries, and numerous shafts and adits along the outcrops of the seams. Most of the pits and quarries have now been filled in, or survive only as shallow hollows in fields. The great majority of the shafts and adits have also been infilled. Except for the Chapel Park area, the remains of the mining industry have not been recorded in detail. There is ample opportunity for a very rewarding project to locate and record all that is left in the form of shafts, adits and surface workings. I have not been able to undertake this myself and have relied on published sources, especially the account of Richard Acworth, and on observations by Barry Hughes in the Somers area.

Owing to the steep dip of the strata, the belt of productive culm seams is quite narrow: for example it is about 440ft wide in the Chapel Park area. The belt is shown on the map, **3.1**, extending relatively uninterrupted across country from west to east, but in reality the culm seams must be offset in many places by the numerous faults that are present in the area; however, these faults do not have a major effect on the displacement of the culm seams. The tendency of the seams to pinch and swell in thickness along their length means that the belt shown on the map is not made up of continuous seams, but instead they come and go so that at some localities there may be very little in the way of culm.

Hartland

Knowledge of the culm mining industry is generally restricted to the Bideford area, but there is also some evidence that culm has been worked around Hartland. There, a seam of carbonaceous material similar to the Paint Seam of the Bideford area was dug and used as a paint pigment. In an 1872 article in the *North Devon Journal* it was reported that Mr John Buse of the Hartland Post Office remembered 'the opening of the Hartland coal mine in the year 1810'. The term 'coal mine' may have been an exaggeration, for the workings seem to have been of limited extent. No anthracite suitable for burning was found, but 'a rich vein of black mineral' was discovered which, when ground with oil, produced a very superior black paint. In the *Hartland Chronicle* Mr R Pearce Chope recorded in 1902 that 'the late Mr Heal, carpenter' had made all his black paint by grinding the culm pigment with oil. The discovery of coal during the construction of the lighthouse at Hartland Point was reported in the *North Devon Herald* of 3 April 1873. Peter Claughton reports that in the early 19th century an unsuccessful attempt was made to work coal at Rosedown near Hartland; paint pigment was worked on a small scale nearby and the name 'Coalpit Lane' is a reminder of the former mine. In 1872 an attempt was made to re-open the Rosedown workings (there were reports in the *North Devon Herald* of 19 September and 7 November 1872).

We now return to the main culm mining district which lies to the west and east of Bideford. Here, as noted above, there are small old pits, adits and shafts, now mostly filled in, scattered along the length of the main outcrop of the culm seams, but the most extensive workings have been in four places: 1. Just east of Bideford, particularly at Chapel Park Mine; 2. at Somers, near Hiscott; 3. at the Tawstock Mine near Fishleigh Barton; and 4. at Hawkridge Wood near Umberleigh.

Following the seams from the coast eastwards to Hawkridge Wood we will explore the various workings. **Please note: the workings are mostly on private land. All disused mine workings are potentially dangerous, and these old culm workings are particularly hazardous. Do not enter any adits or shafts. Not all shafts and adits are recorded, and old shafts in woodland are difficult to locate and therefore especially hazardous.**

In the following account, reference is frequently made to an account of the *Anthracite seams of north Devon* by Richard Acworth, published in the *Journal of the Trevithick Society* in 1991, which contains many details of the workings, and I am most grateful to the Revd Acworth for permission to quote from his work.

Greencliff

The cliffs at Greencliff, about 3 miles west of Bideford, are easily reached by a pleasant short walk of about 700 yards from Greencliff Farm where in 2009 it was possible to park your car (there is limited space, and a small charge was made). In the *Sherborne Mercury* newspaper of 1 December 1800, there is a notice of the sale of an estate called Greencliffe, Abbotsham, on which 'There hath been lately discovered ....a vein of coal which is now working and likely to turn to great advantage'.

The cliffs show steeply inclined shales and sandstones belonging to the uppermost part of the Bideford Formation. A shingle ridge made up of large rounded cobbles up to about a foot across is banked up against the base of the cliff. To the seaward of the shingle ridge

3.3. A view from Greencliff looking southwards towards Cockington Cliff. The limekiln pictured in photo 3.5 is hidden in the small valley in the foreground.

3.4. The culm bed at Greencliff, showing as a nearly vertical black streak in the cliff. The headland is formed of sandstone (the Cornborough Sandstone). The mouth of the old adit was concealed when this photo was taken in August 2009.

is a shore platform cut into the shales and sandstones. There are pleasant views of the coastline (photo 3.3), and on a clear day Lundy Island can be seen on the horizon to the west.

A distinctive feature in the cliffs is a rib of sandstone called the Cornborough Sandstone (page 37). Just to the south of it (above it in order of succession) is a nearly vertical carbonaceous black layer which marks the position of the culm bed (photo 3.4). This seam has been worked in the past by an adit (a horizontal tunnel) driven into the cliff and remains of the adit and the timbers used to support it can still be seen at favourable times. An old shaft has been recorded on the cliff top. When I visited Greencliff in August 2009 the old adit mouth was obscured by material fallen from the cliff.

At the end of the valley, just above the beach, are the remains of a limekiln (photo 3.5) where the anthracite was used as fuel to burn limestone brought in from South Wales. Although generally it is concealed by the pebbles of the storm beach, in suitable conditions traces of a soft coaly culm bed (possibly the Paint Seam) can be seen at the mouth of the stream, about 16ft south of the limekiln. Newell Arber, in his 1911 book on the coast scenery of north Devon, noted

that the Greencliff seam, about 2ft thick, cropped out in the cliff a few yards north of the limekiln, but was '…often more or less obscured by rain-wash, and the coal weathered to a black mud'. He recorded that the seam was worked in 1805, and mentioned that traces of the rubbish heaps of the old workings could still be seen.

The early geologists Adam Sedgwick and Roderick Murchison noted that near the coast the culm beds appeared to have been worked on two lines very near each other. When they visited in the summer of 1836 the miners were extracting the culm for domestic use, but were working one pit sunk on the southern line of culm 'for the sake of the highly carbonaceous and unctuous shale, which runs side by side with a thin band of pure culm, and which is extensively sold as pigment.'

Between the coast at Greencliff and the town of Bideford there seems to be little evidence of any substantial workings on the culm seams, and few shafts or adits have been recorded. However, Peter Claughton records, in his 1993 list of mines, 'Various anthracite and paint pigment workings along line of seams from Greencliff in the west to Bideford in the east'. Trials for anthracite were made in the 19th Century at Abbotsham, about a mile east of Greencliff, and in 1889 Hugh Strong reported that small quantities of coal were said

3.5. The old limekiln at Greencliff, which was fuelled by anthracite from the nearby workings.

to have been found. At Pitt Quarry, Abbotsham (probably SS 420 270) Townshend Hall, writing in 1875, noted that 'a great variety of [fossil] plants are found in the grits adjoining the culm bands'.

Bideford

Workings for culm extended beneath the town of Bideford itself, and several shafts and adits are known. Mining may have begun in the area of Bideford High Street as early as the mid-17th century. Paint pigment was worked from the Union Mineral Black Mine, situated at the top of the High Street in Bideford, near the grounds of the Old Rectory close to the junction with Pitt Lane. Peter Claughton noted that the exact location of the mine is not known, but there is a shaft or shafts in the area of the Rectory grounds at about [SS 451 266]. He noted that the mine was being worked for paint pigment from about 1802 to 1813, possibly on or near the site of an earlier anthracite mine. A third share of the mine was put up for sale at the Auction Mart, London on 31 October 1811, and a copy of the sale particulars is shown in illustration 3.6. The proprietors were Messrs Stocker and Dodge of Bideford.

These particulars give some fascinating details of the mine. It was noted that a shaft 7ft by 8ft and 240 ft deep had been sunk and about 3,000 tons of mineral black had been raised and were lying on the surface. Specimens of the mineral were made available for inspection by interested parties. The sale included a share of the 'Convenient Manufactory contiguous thereto, with Drying Pans, Pulverizing and sifting Apparatus, and every other Requisite for rendering the Article fit for Sale on a large Scale'. There is an unsigned handwritten note on the sale particulars which seems to be a calculation of profit on the venture. The expense of manufacturing per ton was given as £3 10s (£3.50), while the delivered price was £12 per ton, yielding a substantial potential profit of £8 10s (£8.50) per ton.

Peter Claughton recorded an un-named mine for anthracite and paint pigment beneath the upper part of Bideford, apparently

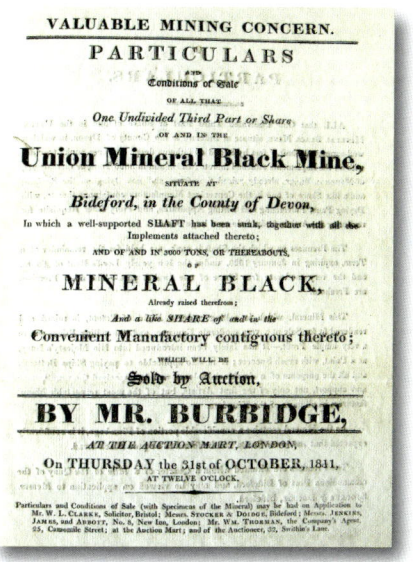

3.6. *Sale particulars for the Union Mineral Black Mine, Bideford, 1811.* Reproduced by courtesy of the Devon Record Office, reference Z18/7.

drained by an adit behind Grenville Nursing Home in Meddon Street. The mine was working at various times in the 18th and 19th century, and possibly earlier. In the 1823 *Pigot's Directory* a company called Cole, How & Co, Mineral Black Manufacturers, was listed with an address in Market Street and again in 1830 with an address in Meddon Street. Of several adits behind Meddon Street one was reported in 1830 as being 1000ft long and may have been the Cole, How & Co working.

A notice in the *Sherborne Mercury* of 2 December 1826 records the discovery of 'a very fine vein of culm of excellent quality' belonging to the Revd William Walter, 'in the pleasure ground adjoining the Vicarage-house at Bideford'. The *North Devon Journal* for 14 September 1827 gives notice of the sale of 'a quantity of culm arising from a mine lately opened in the glebe of the parish of Bideford', but the exact location of this mine is uncertain – possibly it was on Pitt Lane.

East-the-Water to the Chapel Park area

This area is of particular interest because it was the main focus of mining activity for about 150 years from the late 1820s until the final closure of the Chapel Park Mine in 1969. It includes the mines of East-the-Water, Broadstone, Westwood and Chapel Park (the last three are shown on the map, 3.7). Richard Acworth noted that these mines were sometimes worked together and at other times separately. The workings of the various mines in this area were connected, and Felton Vowler notes, for example, that an adit (that shown on the left of the map, 3.9 on page 46) from Westwood Mine reached Barnstaple Street in East-the-Water, a distance of about 1300 yards.

It is this area that was crossed by the new relief road (Manteo Way) that skirts the east side of East-the-Water, and there have also been proposals for housing developments in the Chapel Park area. Consequently the area has been the subject of several engineering studies aimed at identifying the positions of old culm workings, in

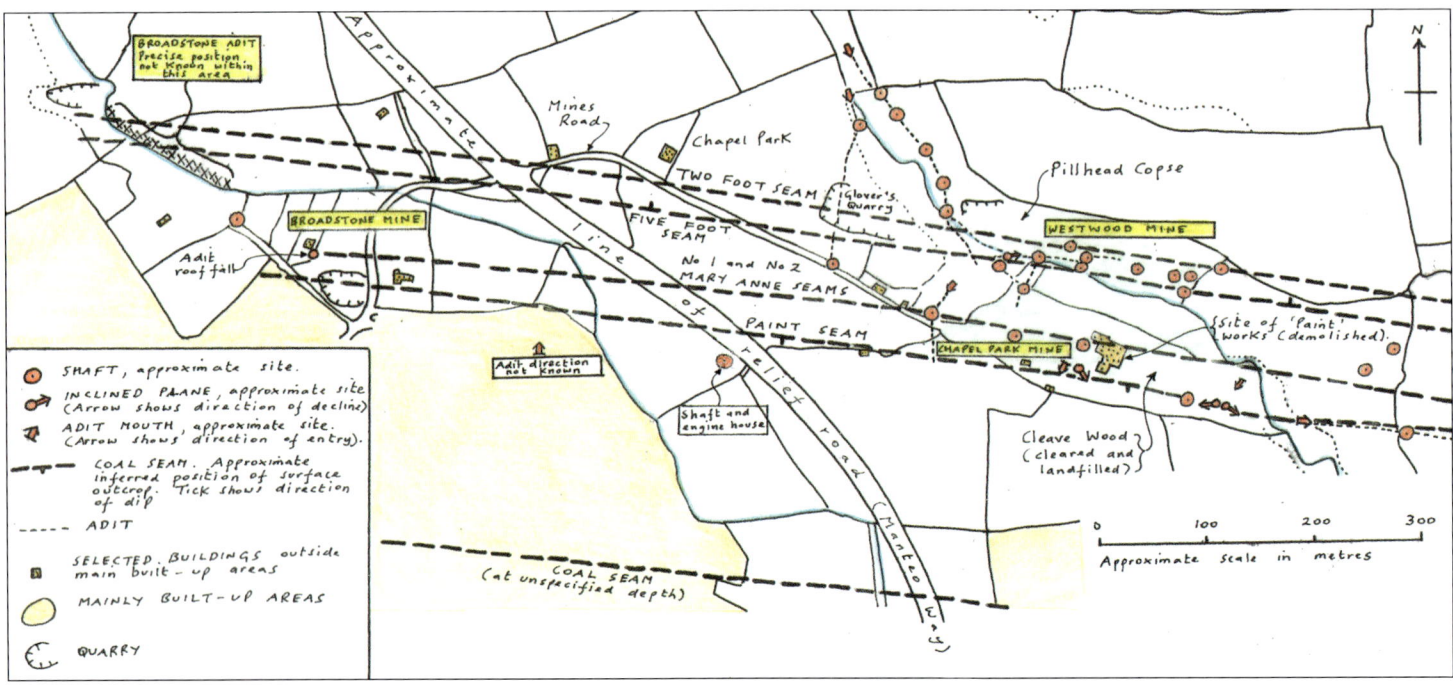

3.7. A sketch-map of the Chapel Park area, East-the-Water, Bideford, showing the interpreted approximate positions of the main mines, the chief culm seams, and the locations of shafts, adits and inclined planes. This drawing is for general illustrative purposes only and is not intended to show the precise positions of mining features. For the location of the area within north Devon, see the map, 3.1. Based, with permission, on Plan No 1 in Report No 1685 by Frederick Sherrell Ltd, April 1996, which contains further details.

particular the locations of shafts and adits. The locations of the Broadstone, Westwood and Chapel Park mines are shown on the map, 3.7, which indicates the position of a selection of shafts and adits and also the conjectural outcrops of the main culm seams. These details have been extracted from reports by Frederick Sherrell Ltd (consulting engineering geologists).

Brian Cubbon notes that in this area culm was found on land owned by Alexander Rowe at Chapple Park. His workings were well established when William Tyeth of Pill Head began mining at West Wood (*North Devon Journal*, 2 March 1827). In 1828, Tyeth tried to raise interest and presumably capital by inviting potential investors to visit the mine (*North Devon Journal*, 31 January 1828). There were proposals to build a tramway or rail road but this did not come about until about 20 years later. The two proprietors Rowe and Tyeth were

rivals and in 1828 were involved in a trespass case, described on page 46, below.

East-the-Water

Workings for Bideford Black and anthracite are known from the centre of East-the-Water on the east bank of the River Torridge. In a notice in the *North Devon Journal* of 18 December 1834 of a house sale in Barnstaple Street, East-the-Water, it was remarked that 'the garden possesses a fine southern aspect, and culm in considerable quantity lies a few feet beneath the surface.'

One of the earliest proprietors of the works was a Mr Pollard who owned mines at East-the-Water (including Chapel Park) in the 1830s. These were sold to the Bideford Anthracite Mining Company in 1846. That company drove an adit [SS 458 265], apparently on the 'North Seam' from just above the quay in East-the-Water (sited in Barnstaple Road, in the car park of the former Ship on Launch Inn – now converted to offices), under Chudleigh Fort, and towards the Broadstone and Chapel Park workings. This adit is no longer visible, being concealed behind outbuildings.

The company also sank a shaft in the area northwest of and below Chudleigh Fort, and installed a pumping engine; apparently a tunnel was driven from this shaft westwards to beneath the bed of the River Torridge. Another adit was driven higher up the hill between the shaft just mentioned and the lower adit; a tramway was laid in the higher adit and descended the hillside to reach the quay by a tramway bridge, or staithe, supported by trestles, which crossed the road and ended in a chute by which culm was delivered directly into ships (illustration 3.8). Most of the traffic was probably by river barges to kilns and potteries. The North Devon Museum Trust collection includes photographs of between 1860-1865 showing ships being loaded at this locality. There were complaints at times of

3.8. *An extract from a painting of about 1862 by an unknown artist, currently on view in the gallery of the Burton Art Gallery and Museum, Bideford. It shows a view of East-the-Water from Upcott Hill. The dark horizontal line in the centre of the picture is the staithe, supported on trestles, by which culm was loaded into boats.* Reproduced by permission of the Burton Art Gallery and Museum, Bideford. © Burton Art Gallery and Museum.

the nuisance from noise and dust where the wagons crossed the road. In the *Exeter Flying Post* for 12 April 1865 it was reported that 'The anthracite mine company at Bideford have resolved upon winding up the concern. Already the old wooden tram bridge across the street has been removed. It had a most unsightly appearance, and the company were to have been indicted at the ensuing quarter sessions for permitting the existence of the tram bridge, which for years has been deemed a nuisance'.

Richard Acworth wrote that production was reported as 10,000 tons in 1851, a figure that is probably exaggerated. The mine was at least 180ft deep in 1856. Both anthracite and Bideford Black were mined in the 1850s and 60s, but subsequently production was concentrated on Bideford Black which was in demand by the Royal Navy for painting ships.

On the painting (illustration 3.8) the horizontal black line in the centre is the tramway bridge or staithe, referred to above, that extended from an adit mouth close to the engine house visible on the hillside. Culm was pushed in tubs along the staithe for loading into barges or ships. As noted above, the structure was dismantled in 1865. Also indistinctly visible on the painting is a steam engine (with a plume of smoke) pulling a line of wagons, close to the northern end of Cross Park, which was the original terminus for the Barnstaple to Bideford railway line (opened in 1855) until the new line to Torrington opened in 1871 and the station was moved to its present site. The goods yards and loading quays remained open at Cross Park, handling considerable traffic until the branch closed.

There are still reminders in more modern times of the former culm mines that extended beneath East-the-Water. For example, a newspaper article in the *Devon & Cornwall Advertiser* of 19 November 1965 reported that a hole some 20ft deep and about 20ft by 15ft across suddenly appeared in the garden of a property at Chudleigh Road, East-the-Water, presumably a collapse into former culm workings.

Broadstone Mine
Broadstone Mine lies between the East-the-Water workings and Chapel Park (see the map, 3.7). Peter Claughton noted that the exact location of the mine is not known, but an adit was reported in the valley to the northwest of Chapel Park. Barry Hughes tells me that adits are still visible in the Littlebrook area of Broadstone. Workings for paint pigment were possibly active in the 1890s. The mine was worked by G and T Pollard in 1891-93 under the name

Broadstone Black Clay Mine. A report by Humphrey Morgans in 1927 mentions a visit to the Broadstone adit in 1926. In a letter of 21 April 1927 from Messrs Hole Seldon & Ward, solicitors of Bideford, it was reported that the No 1 Mary Ann Seam was over 8ft wide in Broadstone Adit, and also in the same adit the walls of No 2 Mary Ann and the 5-foot Seam could be seen. It was noted that in the main adit, No 2 Shaft showed the walls of all the seams mentioned above.

Westwood Mine

The workings of Westwood Mine were situated just to the north of Chapel Park Mine, in the area formerly occupied by Pillhead Copse, as shown on the map, 3.7, although there is some uncertainty about the precise location of the mine. It was opened about 1807, as evidenced by an advertisement in the *Exeter Flying Post* of 25 November 1807 which advertised a lease of '...a Culm Mine, lately discovered, and just opened, on Westwood farm, within the parish of Bideford, Devon. The shaft of this mine has been sunk only 26ft, in sinking which several fine veins of excellent culm have been cut through...'. An advertisement in 1859 (*North Devon Journal* of March 10) offered the lease of culm and mineral paint mines in a field called Cleave on West Wood Farm. Peter Claughton noted that this mine had been worked at an earlier date when an adit had been driven southeast from the valley running north from Chapel Park.

A report in the *North Devon Journal* of 31 January 1828 gives a fascinating account of an incident indicating the rivalry between the two mine owners Alexander Rowe and William Tyeth. This report apparently refers to the Westwood Mine. It appears that '....several thefts and depredations having been committed on the Culm Works belonging to W S Tyeth Esq. of West Wood....' Mr Tyeth set a watch on his mine. He found that the lock which guarded the shaft to the mine had been forced. Two of Mr Tyeth's men descended the shaft to find five of Mr Rowe's men hard at work – their object being to dig a passage to drain large amounts of water from the workings of Mr Rowe into those of Mr Tyeth! The men beat a hasty retreat up the shaft expecting an easy getaway, only to find Mr Tyeth waiting. They escaped, however, but by 20 March 1828 the *North Devon Journal* reported that they had given themselves up for trial. A map (3.9) was produced for the subsequent trial showing details of the trespass, the extent of workings owned by Mr Tyeth, and the boundary with Mr Rowes' workings. An interesting feature of the map is a small

building marked 'Changing House' where Mr Tyeth's men could change their clothes after a shift underground. In this respect Mr Tyeth was in advance of the times, for, as Brian Cubbon points out, it was not until 1872 (following a Royal Commission in 1860) that mine owners employing more than 13 men underground had to provide a 'Dry' for the men.

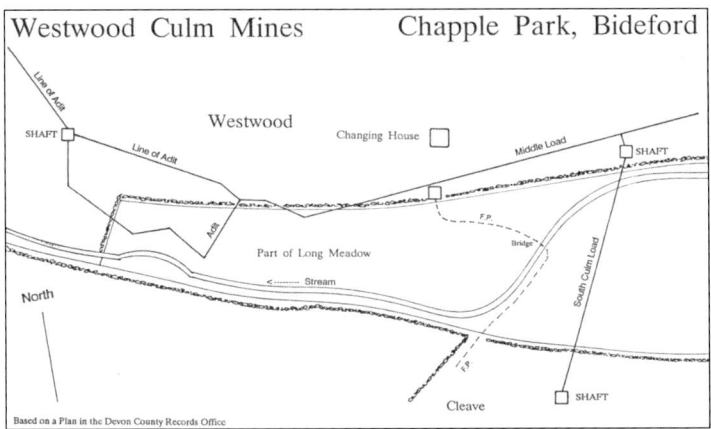

3.9. A plan of the Westwood Mine dating from 1828, drawn up for a court case following a trespass of Mr Rowe's miners on to the property of Mr Tyeth. Based on a map in the Devon Record Office, redrawn by Felton Vowler.

The map (3.9), redrawn from the original, shows three main shafts at Westwood Long Meadow. That on the bottom right of the map leads, according to Felton Vowler, to the 'South Culm Lode' where another shaft had been sunk at the bottom of the valley. This 'lode' dipped to the north towards the stream at the valley bottom and joined the 'Middle Lode'. Felton Vowler notes that the adit shown on the left of the map is the one that reached Barnstaple Street near the Ship on Launch Inn (now closed).

Brian Cubbon writes that in 1842 between thirty and forty men were employed by Mr Tyeth, but by 1868 this had gone down to four men who were then working for Thomas Pollard.

Chapel Park Mine

The best known of all the culm workings are those of the Chapel Park Mine, situated about ¾ mile east of East-the-Water, where mining of Bideford Black continued until 1969. The positions of shafts, adit portals and inclined planes identified during a 1996 survey by Frederick Sherrell Ltd, together with the conjectural positions of the four culm seams worked in this area, are shown on the map, **3.7**. The four seams worked were, from north to south: the 2-foot Seam, the 5-foot Seam, No 1 and No 2 Mary Ann Seams combined, and the Paint Seam. The width of the outcrop belt of all the seams together is about 440ft. It is believed that the Chapel Park Mine, and the Westwood Mine just to the north, were both well established by the 1820s. A mine at 'Chapple Park' was offered for sale by auction on 20 June 1831 (*North Devon Journal* of 9 June 1831), as follows: 'A particularly desirable field called Chapple Park. Situated near the town of Bideford about 8 acres. There is a valuable Culm mine in the field with all necessary shafts, adits etc for working the same & one of the adits is considered essential to the working of the adjacent mine.'

The workings were described by Henry de la Beche in 1839. He noted that, in 1838, culm or anthracite mines were at work about a mile east of Bideford (presumably the Chapel Park area). Workings about '200 fathoms' (1,200ft) long lay above an adit '15 fathoms' (90 ft) deep. The workings were mainly on the 'middle or great anthracite bed' which varied in thickness from 6 inches to 14 feet, with an average of 7 feet. The amount of anthracite produced was about 4,500 tonnes per year for about two years before 1838, but by that year the seam had been worked down to 8 or 10 fathoms (48 to 60ft) which was the limit before it became impossible to drain the workings with the machines then available.

In 1846 the Bideford Anthracite Mining Company took over Chapel Park Mine, and possibly also the Westwood and Broadstone mines; the former owner, Mr Pollard continued as agent until he became bankrupt in 1849. Workings on four anthracite seams produced an output of 500 tons a week. It is reported that by 1856 the workings were 180ft deep and an adit was being driven from near the quay at East-the-Water towards Chapel Park. In 1875 Chapel Park Mine was owned by the Bideford Black Company which continued production until the early 20th century. An Ordnance Survey map of 1887 shows the area west of Cleave Wood occupied by the 'Chapel Park Paint Works' (see the map, **3.10**).

In 1893 Chapel Park Mine was held by G and T Pollard of Bideford together with the Broadstone Mine which was being worked for mineral pigment. A list of mines worked under the Coal Mines Regulation Act in 1896, in a table compiled by Joseph S

Martin, Her Majesty's Inspector for the Southwestern District, showed two mines being worked for 'Black Clay', at Broadstone and at Chapel Park, both in the hands of G and T Pollard. Two underground workers were listed at Broadstone Mine, but none at Chapel Park, suggesting that Chapel Park may not have been actively worked at that date.

About 1920 Chapel Park Mine was re-opened by the Devon Anthracite Company, producing only about 150 tons per year. In 1927 the mining engineer Humphrey M Morgans was commissioned by the company to report on the prospects for anthracite mining on its land. The company had been very active in the 1920s and owned mineral leases all along the known extent of the culm seams. They had prospected the area, opening adits and trenching across the seams at Chapel Park, Somers (near Hiscott), Grabbishaw, Kennacott and Stony Cross. In a letter of 9 July 1926 H W Morgans gives a list of properties visited on 6 July 1926 as: Broadstone adit, Pillhead Copse, Pollard's Quarry (now Glover's Quarry), the Paint Mill, the Paint Seam Adit, Nos 1 and 2 shafts (shown on the map, **3.10**), Stone Wood, Webbery Barton, Stony Cross, Alverdiscott, Summers Hiscot [sic] and Chapelton.

Work had been done on two shafts in the Chapel Park area. A new

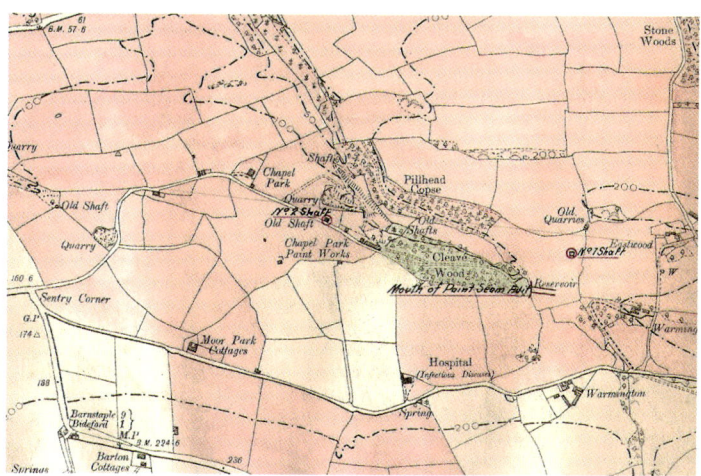

3.10. Part of a 1:2500-scale Ordnance Survey map of 1887 showing the Chapel Park Paint Works and the positions of several shafts. The manuscript additions showing the positions of No. 1 and No. 2 shafts and the mouth of the Paint Seam adit date from the late 1920s. Not to original scale.

shaft [SS 4743 2626], the No 1 Shaft (shown on the map, **3.10**), northwest of Warmington, was sunk to a depth of 96ft. From a depth of 85ft, horizontal roadways were driven due north for 172ft and due south for 166ft. Four coal seams were cut: two north of the shaft and two to the south. The most northerly seam, the 2-foot, was vertical, 15 to 18 inches thick and reported to be 'fairly hard'. Next to the south came the 5-foot seam, dipping north, which in a drive to the east for 40ft varied in thickness between 2ft 10in and 3ft 7in. In the drive to the south of the shaft, two seams, No 1 Mary Ann and No 2 Mary Ann were intersected, separated by only 8ft. The No 1 Mary Ann (the most northerly of the two) was followed in a drive to the west of 34ft where it was 2ft thick, variable, and dipping steeply south. In a drive of 23ft east it was 2ft thick. The most southerly seam cut, the No 2 Mary Ann, was 2ft to 2ft 7in thick and nearly vertical.

The Paint Seam was not intersected, but its projected position was estimated to be about 301ft south of the shaft. In the years 1920-6 the Paint Seam produced 150 tonnes a year of culm, which was crushed and ground to make the Bideford Black paint pigment. In 1927, the Paint Seam was being worked from an adit 1,700ft long, the mouth of which was at the eastern end of Cleave Wood. The adit was driven throughout its length on unfaulted nearly vertical Paint Seam averaging 2ft 6in thick. Humphrey Morgans reported that, before the Cleave Wood adit was opened, the Paint Seam was worked from an adit farther west and there were also workings in Stone Wood about half a mile east of Cleave Wood.

The Devon Anthracite Company also deepened an existing old shaft in the Chapel Park area, No. 2 Shaft [SS 4693 2634], south-south-east of Chapel Park (shown on the map, **3.10**). It was deepened from 110 to 155ft, with at 110ft an adit which provided drainage. The four coal seams had been worked at and above the 110ft level. From a depth of 145ft in the shaft a short drive was made to the east, and then from its end crosscuts were driven due north and due south. The southern drive was abandoned because water from workings broke through. To the north, the 5-foot Seam was cut and was 4ft 4in thick, dipping steeply north at 70 to 75 degrees. Owing to 'heavy ground' the northern drive was not continued far enough northwards to intersect the 2-foot Seam as would be expected.

The mining report by Humphrey Morgans concluded that in the area east of Bideford the amount of workable coal on three of the seams (the two Mary Ann seams and the 5-foot Seam, the 2-foot being regarded as too thin to work) to a depth of only 600 yards was 20

million tons. It was recommended that a colliery with two shafts be established in Cleave Wood, connected to a quay on the river at Bideford by aerial ropeway or railway. These grand plans never came to fruition, of course, possibly because of lack of finance in the 1929-30 slump. Plans to develop the anthracite reserves were shelved in favour of production of the mineral pigment. Bideford Black Ltd was formed to take over the mineral black manufacturing business of Devon Anthracite Ltd and was registered on 20 December 1928. The authorised capital was £100,000, divided into 400,000 shares of 5s each.

The prospectus issued by Bideford Black Ltd in January 1929 gives a fascinating snapshot of the industry at that date. It noted that the company had the only source of commercial mineral black pigment in the country. It was estimated that to a depth of 600 yards the mineral black seam would yield 5 million tons of unrefined material. The property covered much of the total outcrop of the culm seams over a straight line distance of about 12 miles east and west of Bideford, an area of about 5,500 acres. The property was held for fifty years from December 1925 and other dates at 1s per ton on carbon black, 6d per ton on anthracite, 1s per ton on wayleaves, and various dead rents amounting to about £250-300 per year. 'Carbon black' was at that date being extracted from an adit running east; a new shaft (No 1 on the map, **3.10**) had been sunk to a depth of 96ft and an old one was being re-opened (No 2 on the map **3.10**). The company had plans to build a new treatment plant with an output of 8,000 tons per year.

The new company sank an inclined shaft at Chapel Park on the Paint Seam at the eastern end of Cleave Wood (see the map, **3.10**). The narrow gauge tramway ran along an incline which was up to 400ft long, and the tubs loaded with culm were winched up the incline, the mouth of which is shown in photo **3.11**. From the mouth of the adit (which was fitted with a door), the tramway led on log sleepers to the processing plant which was constructed on the hillside in Cleave Wood (see page **55**). The purpose of the curious candle fixed to the front of a mine tub, visible in photo **3.11** is explained by Brian Cubbon. A lighted candle was fixed to the front of the tub on top of the locking bolt to guide the winch man at the top of the incline since he could not see the tub at the start of the ascent. Communication between the surface and the underground workings was by a series of bell codes.

Unfortunately, problems arose with the process which attempted to

upgrade the carbon content of the pigment by chemical means (see page 56), and Bideford Black Ltd went out of business in December 1935. The works were taken over by Bideford Black Pigments Ltd, a private company under the direction of Mr H (Howard) St L Cookes, a local man. The chemical process was abandoned and the Paint Seam material was simply dried and ground and sold as a powder for use in paints, with rubber and for colouring cement tiles.

During the last few years of the industry at Chapel Park only a few people were employed. The owner, Howard Cookes, occupied a small office on site with a secretary. Three men worked below ground and another three were involved with drying, processing and packing the Bideford Black. The *Mining Journal Annual Review* of May 1954 noted that small-scale mining continued at the mine.

The Chapel Park Mine closed in 1969 when the market was taken over by synthetic chemicals. Richard Acworth reported that the Chapel Park site was sold as a scrap yard in 1972, the plant was demolished and the shafts had been filled. He wrote that in a field nearby the remains of a winding engine house could still be seen (in 1991). A deep adit still drained the main workings into the valley below Chapel Park, and another adit below Chapel Park was open for 20 or 30 ft, but there was a fall at the position of the Mary Ann seams.

Today little can be seen at the Chapel Park site. There are few surface remains and no buildings survive. The site has been largely landfilled. There is no trace of the adit at the eastern end of what was

3.11. A photo, probably taken at Chapel Park Mine, showing men pushing a mine tub loaded with raw culm. The wheels are double-flanged and there is an end-tipping door on the wagon. Photo from North Devon Museum Trust Collection.

Left: *3.12. Spoil heaps of anthracitic material at the eastern end of what was Cleave Wood, Chapel Park Mine, in 2009.*

Below: *3.13. Detail of anthracitic material on spoil heaps at Chapel Park Mine in 2009.*

Cleave Wood (about SS 4733 2619), but there are several spoil heaps of anthracitic material (photos **3.12** and **3.13**). The one adit surviving is at the western end of the site, just north of where the track descends to the valley. Here, in the valley side is an adit (about SS 4706 2628), now partly rubbish filled; it is probably the one referred to by Richard Acworth as being open for 20 to 30ft in 1991. South of the workings, an old wall at a field boundary (about SS 4692 2625), not far east of the new relief road (Manteo Way) may be the remains of a mine manager's house, according to information from Barry Hughes. This structure is shown on the 1840 Tithe Map and on later Ordnance Survey maps.

Immediately west of Manteo Way are the remains, at about SS 4635 2626, of a building which may have been a winding engine house, with an adjacent infilled shaft (the location is shown on the map, **3.7**). Nothing is shown on the 1840 Tithe Map at this site, but a building is shown on the 1888 and 1904 1:2500-scale Ordnance Survey maps (although its function is not identified). The 1888 map marks an 'Old Shaft' just southwest of the building.

Felton Vowler visited the Paint Seam adit in the 1980s and found that there were still two tubs, similar to those pictured in photo **3.11**, sitting on the tramway at the bottom of the incline; they were very rusty and too dangerous to remove. The tubs ran on 1ft 9½-inch track and the wheels were double-flanged, similar to the slate wagons of North Wales.

Stone Wood; Long Down Wood; Webbery; Stony Cross

Workings for culm continued east of Chapel Park (see the map, **3.1**). Richard Acworth noted that the workings extended under Eastwood to Stone Wood and Long Down Wood (now cut down), and in 1991 an adit was open in Stone Wood. A 1927 letter from Hole Seldon and Ward noted that 'At Stone Wood can be seen the walls of No 1 Mary Ann and numerous caved-in pits on the 5-foot and 2-foot seams'. H St L Cookes recorded 'signs of old shafts or workings' in Long Down Wood and Southern Down Wood. The seams continue to Webbery and Stony Cross, although there do not seem to have been any significant workings. Richard Acworth, writing in 1991, noted that there were extensive workings in Alverdiscott Parish from Webbery, where the sites of two shafts could still be seen, continuing through Webbery Woods where workings were visible until the woods were cut down and the workings bulldozed before re-planting, under Bulworthy to just north of Stony Cross where the site of an adit could

be seen. Peter Claughton, in his 1993 list of mines, recorded an unnamed anthracite mine [about SS 505 258] in Webbery Wood, Alverdiscott Parish, where there were shallow shafts and open workings, with further workings along the seam to east and west.

Workings were reported by Richard Acworth between Alverdiscott Church and East Woodlands, possibly on the most southerly (?Paint) seam (see the map, **3.1**).

West Kennacott to Grabbishaw

East of Kennacott to Hawkridge the culm seams lie apparently within the Bideford Formation outcrop but may be in unmappable synclinal areas of basal Bude Formation beds.

Leases of 1759 and 1760 refer to mineral workings west of Hiscott (the exact location is not known), at West Kennacott, Grabbishaw and Punchard's Down (the location of the latter place is uncertain, but it seems to have adjoined West Kennacott). Richard Acworth noted that '...there have been workings in Grabbishaw Wood within living memory, and...anthracite was encountered when they sank a well at the top of the hill behind the house'. The seams worked at Grabbishaw extend from Bartridge Hill through Kennacott and Grabbishaw to south of Hiscott and on to Somers.

Somers, near Hiscott

The next area with extensive remains of mining activity is in the Somers area, near Hiscott (see the map, **3.1**). Richard Acworth wrote that: 'The seam runs under the hill on which the house [of Somers] stands, and there are considerable remains in both valleys. On the north side there was apparently a deep adit a quarter of a mile long which ran south under the hill and which was worked for many years before closing in about 1838. The site of a shaft can also be seen half way up the hill on that side, and the sites of several shafts can be seen near the houses at the top. But the most extensive remains are in the valley to the south of the house, where a succession of shafts and adits can be identified. One adit is more or less open, oozing black mud, while another, which people alive have explored, is now bricked up with a pipe for drainage. On the hillside above, a badger's set turns up lumps of pure anthracite, some of which I myself have collected and burned. The site of another shaft can be seen further up the little valley. These extensive mines were worked in the 18th and through the first half of the 19th Century. In 1847 they were taken up by the North Devon Coal and Culm Company.....the

Devon Anthracite Company...took a lease of the mineral rights for fifty years from 1925...'

Barry Hughes made some valuable preliminary observations on the remains of the culm mining industry around Somers in 2007, but he notes that a detailed survey would have to be carried out to confirm his findings. The Somers area also has considerable wildlife interest, and there is a network of paths and glades designed to create habitats for wildlife. The positions of the features described below are based on a sketch map kindly provided by Barry Hughes, and are given as a rough guide only. There is a line of shafts that trends approximately east-west. At the most westerly point (about SS 555 257), spoil tips and shafts are present on the top of the hill, and in the tennis court west of the house there is a pit on the line of the shafts. Just east of the house, there are pits and shafts in the woods (very approximately SS 557 257). Continuing eastwards, in the valley bottom is a pit indicating the likely former site (possibly SS 558 257) of a 30ft water wheel used for pumping water out of the mine. It was probably connected by a line of flat rods which passed under the road to a mine adit. The wheel was powered by water from a leat which left the stream about 500 yards farther west, at about SS 554 255. Barry Hughes notes that at this point there was a pond and possibly a weir which has now been washed away. The water flowed eastwards along the leat, following the contour line, and then passed over the road to the water wheel; its former course is now part of a trackway. Away from the main line of shafts, in the area southwest of the house, Barry Hughes recorded the approximate site of a shaft in an orchard just east of a hide on the edge of Bluebell Wood. South of there, not far from where the leat leaves the stream, is an adit which drained the mine and was possibly also used for access.

Peter Claughton gave the location of adits adjacent to Somers as SS 557 256, probably the site of the workings of the North Devon Coal and Culm Company in 1847.

Rather surprisingly the 1:2500-scale Ordnance Survey map of 1888 shows no shafts or other evidence of mining in the area around Somers, except for a trackway following the former course of the leat described above. In 1888 the area was known as Hiscott Down and the woods around the house are marked Hiscott Down Wood, but on the 1904 1:2500-scale Ordnance Survey map the names had changed to Somer's Hiscott and Somer's Hiscott Wood.

A culm mine at Hiscott was offered for sale in the *Sherborne Mercury* for 4 November 1799, as follows: 'Culm Mine to be sold for term of 21 years, or let in shares. All that valuable culm mine on a Messuage Tenement called Hiscott in the Parish of Tawstock now in full work and in high perfection and carried on by Mr James Burrow the sole proprietor. The works are about 5 miles from Barnstaple and 7 miles from Torrington, where are limekilns in or near these towns. There are also many other limekilns in the several parishes of South Molton, Filleigh, Buckland, Swimbridge, Landkey and North Tawton, all customers of the said works, for burning lime. The vein at present is 4ft and upwards in breadth and a depth from the surface of about 25 fathoms, the veins run from east to west. Any gentlemen adventurers willing to purchase the above mine, let them apply to the said Mr Burrow residing on the premises who will treat the same'.

Tawstock Mine

Proceeding eastwards, the next location of interest is the Tawstock Anthracite (or Coal) Mine. Richard Acworth noted that the site of the mine, which had been lost, was relocated as being 'in the fields between the road and the railway just north-east of Fishleigh Barton' [SS 584 249] where the sites of several adits and a possible shaft can be seen. Peter Claughton gave the location of the 'Tawstock Culm Works' as approximately [SS 585 251] 300 yards north of Fishleigh Barton (see the map, 3.1).

Working seems to have begun around 1760. The mine was mentioned in the newspaper report of 1793 referred to by Richard Polwhele in his *History of Devonshire* (1797), as follows: 'It is with peculiar pleasure we announce to the public, that an excellent coal-mine hath lately been discovered at Tawstock, in Devon...several veins having been explored in different directions, the coals of which prove to be extremely good, and burn remarkably clear. This circumstance is likely to turn out highly beneficial to the city [of Exeter], as the line of the intended public canal from Exeter to Barnstaple will run within a few yards of the said coal-mine'. There is a report that there were two seams about 9ft thick producing 900 bushels per week, but abandoned in 1800 owing to difficulty in keeping out water.

An advertisement in the *Sherborne Mercury* of 7 November 1796 asked for '...a Captain or Foreman capable of managing the water engines and mines of the said works [Tawstock Culm Works]. An attentive man whose character and abilities will bear the strictest enquiry will meet with proper encouragement by applying to Mr John Pitt, Clerk of the said works, Barnstaple'.

Hawkridge Wood

The most easterly of the significant culm workings are in Hawkridge Wood (map **3.1**), on the north side of the valley of the River Taw, north of Umberleigh House, and seem to have been for anthracite, with no record of pigment workings. Richard Acworth noted that the workings were on quite an extensive scale, but there are few details available. There is a notice in the *Mining Journal* of the mines being in work in 1846. Richard Acworth gave the following account in 1991: 'There are quite extensive remains in the woods, including an open adit near the River Taw (now a Nature Reserve for the bats which winter in the mine) and the sites of several collapsed adits along the same river bank. Half way up a side valley that comes down through the woods is another open adit, and the sites of at least four shafts can be seen at the top of the wood. One of these is open and in quite good repair; it is said to be about 150 feet deep, but was filled with water when I saw it'. In the Geological Survey Ilfracombe and Barnstaple Memoir, Eric Edmonds and others noted that '.... one shaft in Hawkridge Wood [SS 6054 2514] was reputedly sunk to a depth of about 31m [102ft]'.

Peter Claughton gave the location of adits and shafts in Hawkridge Wood, part of extensive anthracite workings of unknown date, as [SS 605 251]. He noted that the Hawkridge Wood workings seem to have been opened up at various dates throughout the 19th century; little is known of the workings but in 1874 a sale notice stated that they had been quite successful (Devon Record Office, 62/9/2 box 3/1).

An advertisement was placed in the *Sherborne Mercury* on 28 March 1796, seeking workmen for the mine, as follows: 'Wanted by the proprietors of Hawkridge Wood, situated in the parish of Chittlehampton a sufficient number of men to drive an adit of about 300 fathoms. Attendance will be given on Thursday the 14th of April next at the Sun near Umberleigh Bridge to receive the proposals of such persons as may be willing to engage in driving the said adit and in such other employ as may be thought necessary in carrying out the said works. Wanted also a man well qualified and recommended to superintend the said works, and to whom proper encouragement will be given. In the meantime for further information application may be made to Mr John Isaac in Newport near Barnstaple, Devon'.

HOW THE CULM WAS WORKED

In the early days, the culm was dug out using pick and shovel, but by the late 1920s air drills were in use. Dorothy Cleaver and others give a vivid account of mining methods and conditions just before the Second World War, recalled by a miner William Spry who started work in the Chapel Park mine in 1938, aged about 16. He reported walking down to the face, first down a very steep incline for about fifty feet or more, then for about a quarter of a mile to the main working area. The Bideford Black seam was dug out with air drills, generally about 6 to 7 feet of the face being dug in one shift. The miners then selected fir timbers on the surface to support the workings. The pigment was loaded into tubs running on the rails of a narrow gauge tramway, which were winched to the surface by somewhat erratic electrically-operated cable winding machinery. On a good day they would load three or four tubs. The miners carried carbide lamps for illumination, and these were notoriously prone to going out and plunging the miners into darkness if struck against something, or dropped.

Unlike the coal seams of typical coalfields, where the seams are generally more or less flat-lying and can be worked along horizontal tunnels, the Bideford seams are very steeply dipping (at 70° or greater), or even vertical. This presented problems of working, and the general method adopted was to work the anthracite seams in the manner of mineral lodes. A shaft was sunk, either vertically to the seam, or on the incline down the seam. From the shaft horizontal tunnels were driven out along the seam at different levels. From these levels, the coal was dug out from above and below, a method called stoping. Eventually the workings from one level would connect with those from higher or lower levels forming a void of worked-out coal called a stope. The walls of the stope were supported by pillars of unworked coal or, where all the coal had been removed, by timber props. Peter Claughton noted that it is doubtful if any of the earlier workings ventured below adit level except in exceptionally dry weather.

The Bideford Black or Paint Seam was worked somewhat differently to the anthracite seams, at least during the 20th century. The method adopted was to drive inclined planes or 'declines' down the seam. From the decline, horizontal 'roads' were driven along the seam at close intervals, leaving an unworked portion of the Paint Seam about 2 to 3ft thick between each 'road'. Donald Gill, writing in an unpublished report in 1941, noted that the walls of the seams were highly polished and strong, but flaked off and required close timbering. The tunnels were supported by timber props put into place as the working proceeded. In order to maximise production during the Second World War, the whole of the Paint Seam was removed, resulting in subsidence at ground level which affected the Bideford Black treatment plant.

THE CULM MINERS

Little is known of the men who actually worked the mines, but information from the Census for the snapshot of time from 1851 to 1891 indicates that they were they were mostly local men, with a scattering of Cornishmen. However, Dorothy Cleaver and others noted in their 1994 account that during the time of the great depression in the 1920s unemployed South Wales coal miners came to the Bideford mines looking for work. The number of men working in the mines fluctuated in the 19th century. In 1842 between 30 and 40 men were at work. Information gleaned by Dorothy Cleaver and her co-authors from Bideford census returns

Census year	1851	1861	1871	1881	1891
Miners (not differentiated)			8		
Culm/coal miners	19	12	2		
Paint miners		3		2	
Paint miner & agricultural labourer	1				
Engineers/engine drivers	1	2			
Mine captains		1			
Labourers	3				
Accountants	2 (one is George Pollard)				
Proprietors	1 (John G Maxwell)	1 (Thomas Pollard)	1 (Thomas Pollard)	2 (Thomas Pollard & George Pollard)	1 (Thomas Pollard)
Totals	28 All local men except for 5 from Cornwall and 1 from Hunts.	20 All local men except for 4 from Cornwall, 1 from Ireland and 1 from Wales.	11 All local men except for 3 from Cornwall.	4 2 local men; 1 from Cornwall; 1 from Ireland.	1

3.14. Workers in the culm mines, based on information in the census returns for 1851 to 1891, given in Cleaver and others (1994).

gives the number of men employed in various roles in the mining industry (see the table, 3.14). Their account has an appendix giving the names, ages and addresses of those employed in the culm industry between 1851 and 1891. However this is of course a partial picture since it refers only to the Bideford area. The numbers of men involved in the culm industry during this period was fairly small. The figures show a steady decline in the number of men employed from 28 in 1851 to only 1 in 1891. The only person recorded as involved with the industry in 1891 was in fact the proprietor Thomas Pollard who, if the census return is accurate, employed no workers at that time. Even in 1881 only two paint miners were employed by the proprietors Thomas and George Pollard.

The Pollard family

It can be seen from the census returns (3.14) that the Pollard family played an important role in the culm industry during the period recorded. Dorothy Cleaver and others, writing in 1994, have summarised their association, and the following paragraphs are based on their account. They recorded that Thomas Pollard was a Cornishman from Kea, near Truro and may have gained experience in the local tin mines. He is thought to have come to Devon in the early 1820s as a mine captain or agent for the Mineral Black Company, and is listed as such in the 1841 census return. He sold his interest in the Chapel Park Mine in 1846 to the recently formed Bideford Anthracite Mining Company. The company was taken over in 1848 by Mr John Maxwell (*North Devon Journal* for 18 January 1849). His name is listed in the census return for 1851 as a proprietor - see the table, 3.14. This began a period of financial problems and confrontations between Thomas Pollard and the directors of the company; the details are not known, but Pollard was recorded as bankrupt in 1849 (*North Devon Journal* for 11 January 1849) and on 9 August of the same year a public notice in the *North Devon Journal* appeared stating that he was no longer an agent for the company. However he seems to have recovered remarkably well from these difficulties, for in 1861 he was listed in the census as a Mineral Black Manufacturer with four men in his employ. He continued to thrive in local affairs and became Mayor of Bideford in 1879-80. When he died in 1886 aged 80 there was much public demonstration of respect and it is reported that ships in the port of Bideford lowered their flags to half mast.

Thomas Pollard had two sons: George, born in 1833 and Thomas born in 1837. George worked with his father in the family business and also had shipping interests. He was also involved in local affairs and became a councillor, but he was somewhat in the shadow of his father and later his younger brother. Thomas the younger went to Australia as a young man but returned to Bideford aged 33 and bought Chapel Park House and became involved in the family business. He is recorded as 'Quarry Overseer and Mineral Paint Manufacturer' in the 1871 census return. Like his father he was much involved in local affairs and in 1920-21 was Mayor of Bideford. In recognition of his service to the town he was elected a Freeman of Bideford just before his death in 1924.

Accidents

The work carried the usual hazards of any underground mining operation – roof and wall falls, flooding, and poor ventilation, and consequently accidents were not uncommon. Methane or 'firedamp', the highly flammable gas which caused so many disastrous explosions in coal mines, especially before the invention of the Davy lamp, does not seem to have been a problem in the culm mines.

Alison Grant and Peter Christie writing in *The Book of Bideford* (1987) note the dangers of working in the mines. They recorded in the Bideford registers a 1655 entry referring to 'John Boorges, Collier, found deade in the Culme worke'. They note that the graves of men who died in mine accidents can be seen in the Old Town Cemetery. Brian Cubbon writes that between 1827 and 1868 the local press carried reports of seventeen accidents in the culm mines involving over twenty men, with nine leading to deaths. Some details are given below.

A report in the *North Devon Journal* for 18 June 1829 illustrated in somewhat gruesome detail the hazards of working the mines: 'On Thursday last a wall-plate in the mine of W S Tyeth Esq gave way and a portion of the mine fell in and buried two of the miners named William Harris and William Wills; the most prompt exertions were made to afford them relief and the latter was taken up alive, though severely injured, but the former had died beneath the extreme pressure on his chest. He was heard for about 10 minutes before, calling upon God for mercy. He was a native of Redruth in Cornwall and has left a widow and three children in the most destitute circumstances. Several benevolent ladies of this town are locally employed in promoting a subscription for their support'.

In February 1840 a Mr William Gorrill fell 30ft when climbing down into a culm mine (not identified) and was severely injured. An accident at the Chapel Park Mine was reported in the *Mining Journal* for September 1842. It was reported that a drift had been carried beneath an unknown old shaft, a large part of which collapsed on to the miners, one of whom, a Bartholomew Pickard, was killed. His body was only retrieved after eight days and nights digging.

Two further accidents were referred to by Dorothy Cleaver and others. The first occurred in August 1843 when a man named Harris fell down a shaft, hitting buckets on the way down, severing a leg. In the second incident, in December 1849, a young man called William had a miraculous escape when he plunged 130ft down a mine shaft into 7ft of water and escaped with only a few bruises. In 1846 it was reported in the *North Devon Journal* of 19 February that a miner had survived a fall down a 100ft shaft. In 1857 a miner died of tetanus after his leg had been crushed by a falling boulder (*Bideford Gazette*, 20 October 1857). Richard Acworth noted that working of culm from tunnels beneath the River Torridge was said to have ended after an influx of water which killed several miners.

HOW ANTHRACITE AND BIDEFORD BLACK WERE USED AND PROCESSED

Anthracite

The main use of anthracite was as a fuel, and it was widely employed in kilns to burn limestone for agricultural use and also for drying malt. Owing to its crushed nature it was not suitable for household use for which bituminous coals were imported, mainly from Wales. Peter Claughton has noted that in 1849 representations were made, without success, to the Great Western Railway to use the Devon anthracite (*North Devon Journal* for 22 February 1849). He also records that north Devon anthracite was successfully used for smelting lead at Combe Martin for a short time in the late 1840s and early 1850s. The authors of the 1987 *Book of Bideford* note that an agricultural writer of the 1790s reported that not only was limestone brought by sea from Wales to burn in kilns to produce lime, but also culm from Wales to fire the kilns. Whether this was in preference to local culm or for convenience is uncertain.

Bideford Black

Brian Cubbon has listed a wide range of applications for Bideford Black which, apart from the main use as paint, included varnishes, printing ink, coloured paper, stove polish for cooking ranges, cement products, bricks, tiles, linoleum, tyres and tarpaulins. During the Second World War there was a great demand for the Bideford Black pigment for use in blackout and camouflage paint. It has even found a use in womens' makeup, and Dorothy Cleaver and others noted that by the 1950s considerable amounts of the pigment were being used by Max Factor in cosmetics, especially mascara.

In 1800 Bideford Black was in use as a pigment for painting ships of the Royal Navy. The Lysons, in their book *Magna Britannia*, wrote that it was used at Plymouth naval dockyard in 1822, and had also been sent to Chatham dockyard for the same purpose. Even in 1859 the Royal Navy was still advertising in the *North Devon Journal* (4 August) for a supply of Bideford Black for painting warships.

The particulars of the sale of the Union Black Mineral Black Mine in 1811, referred to on page **43**, and pictured in illustration **3.6**, are of considerable interest. The sale was taking place at a time when the Royal Navy was heavily engaged in the struggle against Napoleon. The sale particulars note that the mineral black had '....lately been introduced into His Majesty's Navy, as a Paint, with much Success; it is also applicable to paying [painting] Ships Bottoms, and all the purposes of a dry black Colour; and has received the countenance and support, not only of the first Artists, but of the most respectable Manufacturers in all parts of the Kingdom'. The sale particulars are also interesting because they give some hints about the manufacturing process for mineral black at this time. They note the 'Convenient Manufactory contiguous thereto, with Drying Pans, Pulverizing and sifting Apparatus, and every other Requisite for rendering the Article fit for Sale on a large Scale'. The process thus seems to have involved first, drying the raw mined product; secondly 'pulverising' [grinding?] it; and lastly sieving it to remove any remaining coarse material.

Cole, How and Company were described in Pigot's Directory as 'Mineral Black Paint Manufacturers', with premises in Market Street and by 1830 in Meddon Street.

Production from a mine in Bideford, the location of which is uncertain, was included in figures for 'Ochre and Umber' in the *Mineral Statistics* report for 1887. A total of 3,204 tons was raised, realising £3,845 but these figures include two works in Ashburton as well as the Bideford mine.

In a report of 1927 by Humphrey Morgans for Devon Anthracite Ltd, the current plant for production of Bideford Black was described as an edge runner grinder with a capacity of one ton per day, passing a 90-mesh sieve. There was an elevator and screen, a 5 horse-power paraffin oil engine and anthracite-burning drying stoves. It was recommended that a plant was built to crush the material finer and for concentrating it (this plant was built, and is described below). The output in the seven years ended June 1926 was at the rate of 150 tons per year. The cost of preparing the Bideford Black for sale was £2 15s per ton and the material fetched £5 per ton 'as now prepared in a somewhat crude manner'.

Experiments were made to improve the quality of the final product. In 1928 it was reported (by H M Langton in the *Industrial Chemist* of August that year) that experiments had increased the carbon content from 33% (the amount in the standard Bideford Black product) to around 70% on average and the fineness from around 90 mesh to around 200 mesh, to produce a more saleable product. Following the apparent success of these experiments, a dressing plant for Bideford Black was erected in about 1929 by Bideford Black Ltd at the Chapel Park Mine. The original buildings of the Chapel Park Mine (the 'Chapel Park Paint Works' which were located at the eastern end of Mines Road (see the extract from the 1887 Ordnance Survey map, **3.10**) were probably demolished after construction of the new treatment plant.

Details of the new treatment plant are given in an article by Humphrey M Morgans in the *Transactions of the Institution of Mining and Metallurgy* for 1930. The following is a summary of the various stages of treatment of the raw material. It was brought from underground in wagons directly to the dressing plant which was built on the side of a hill in Cleave Wood (photo **3.15**) so that movement of the material through the various processes was helped by gravity. The chief aim of the process was to reduce the ash content (normally about 70%) in the finished product to proportions most suitable for a saleable product. The first step was to grind the material wet in a Hardinge mill to 40 mesh size. The mill was lined with silex stone and 'Danish pebbles' (pebbles of superior hardness, toughness, and uniformity, found on the shores of Greenland and marketed through Denmark) were used for grinding. From there the pulp was delivered to a Dorr classifier which separated the coarse solids and returned them to the feed to the mill. The fine material passed to a series of flotation units; in the first ten the pulp was treated in series and then passed to the last six units for further concentration. The reagents used in the cells were creosote, cresylic

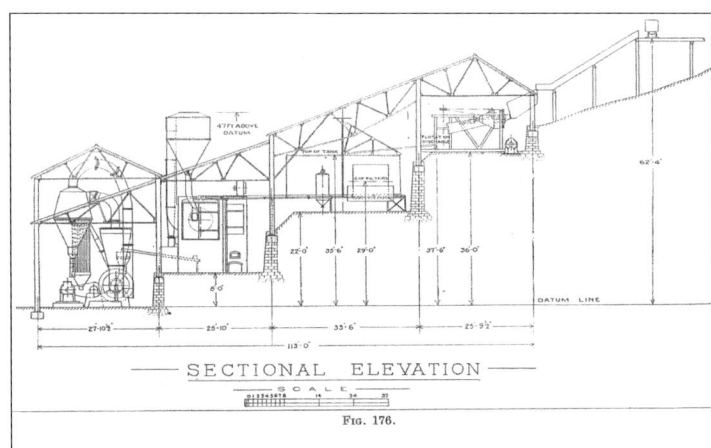

SECTIONAL ELEVATION

Fig. 176.

Above left: *3.15. A photo of about 1933 showing the dressing plant for Bideford Black, built on the hillside in Cleave Wood.* Photo from North Devon Museum Trust Collection. Above right: *3.16. A cross-section through the Bideford Black plant pictured in illustration 3.15.*

acid, pine oil and xanthate. From the flotation cells the concentrate flowed to a Dorr thickener and the thickened pulp was drawn from the bottom of this tank to Oliver filters. The cake produced on the filters was delivered by rubber conveyor belt to a cylindrical drier. The plant was designed to produce 25cwt per hour of finished Bideford Black of a fineness of 300 mesh. It had not yet been established how many tons of raw material were needed to produce one ton of Bideford Black. The water supply was drawn from old mine workings at about 120 gallons per minute. The plant was driven by electric motors. The illustration, 3.16, shows a cross-section through the plant which can be compared to photo 3.15.

The plant cost £14,000, a substantial sum in 1929, equivalent to about £470,000 in today's money, and was built by local contractors Glover and Son. However, the plant stood idle all the summer of 1930 owing to lack of capital. Unfortunately it became apparent that there were problems with the chemical process and the company went into liquidation in 1935. The works were taken over by Bideford Black Pigments Ltd; the chemical process was abandoned and the Paint Seam material was simply dried and ground and sold as a powder for use in paints, with rubber and for colouring cement tiles. Working with this dried product could be a very messy business. Dorothy Cleaver and others report that some of the crushed pigment, in dry powder form, was packed in to

hundredweight sacks to be transported in trucks to the goods yard at the nearby Southern Railway station. A former mine employee, quoted by Felton Vowler in 2004, said: 'I used to work at the Paint Mines, East-the-Water, where I loaded pigment into half-hundredweight sacks onto a Foden lorry to take it to the railway goods yard near the river. When I got there the railway staff refused to handle the sacks because they said they were too dirty, so I ended up putting them into the railway vans myself!' Some of the pigment was packed into boxes weighing a pound or half a pound, then put into tea chests for export to many different parts of the world.

The illustration, 3.17, shows the Bideford Black works at some date between 1945 and 1948. The building on the far right is the highest part of the Bideford Black dressing plant pictured in photo 3.15. The tramway leads from an inclined shaft behind the photographer; a winding cable can be seen on its left. The white building directly ahead is the incline winding house which used electric motors. Tubs of raw culm were tipped into the plant. The building with the curved roof is the modest Works Office; just to the right of it are farm buildings at the end of Mines Road. All of this is now gone.

Three grades of product were produced (see the *Industrial Chemist* for February 1933). *Fillablack* consisted of the material from the seam, simply dried, crushed and ground and was essentially the old

'Bideford Black'. Because of its general properties and its specific gravity (2.14) it was suitable as a general cheap filler to be used where a dark-coloured product was needed, for example to replace slate dust and barytes. *Biddiblack* was obtained from the concentrating plant. It was sold as a pigment for use in the manufacture of paints, cement, tiles, paper and ink. The product *Bettablack* was considered superior even to Biddiblack as a rubber reinforcing agent. The method of production was simpler than that of Biddiblack, for the flotation process was dispensed with. There also seems to have been a product called *Jetablack*, the nature and use of which I cannot determine. Dorothy Cleaver and others recorded that in the 1920s a local chemist, Mr G Phillips, set up a small laboratory at his home in Barnstaple Street, Bideford, to manufacture soap and disinfectant from the pitch-like material recovered from the workings, but its strong carbolic small reduced its sales appeal. Fillablack Road and Biddiblack Way, located in new housing developments in Bideford East-the-Water, are named after these products as a reminder of the industrial heritage of the town.

A 1941 report by Donald Gill noted that output before the

Second World War was about 300 tons per year, but in 1941 was 600 tons per month because of the wartime demand for blackout and camouflage paint. However, because the whole of the seam was extracted during the Second World War to increase output, there was surface subsidence which caused damage to the mill at Chapel Park. The material was hand-sorted and treated in a Scott rotary drying furnace. It was then ground so that 99% passed a 200 mesh sieve screen. H G Dines, writing in 1947, noted that 'The present Bideford Black Pigments Ltd mine appears to have raised about 300 tons a year in pre-war days, for pigment. During the war, the material was in demand for blackout purposes and attempts were made to increase the yield to over 300 tons per month, but with what success is not recorded'.

Production of Bideford Black by Bideford Black Pigments Ltd continued until 1969. The works were forced to close for several reasons: there was competition from low-priced synthetic carbon black; cheap black filling was no longer advantageous in many manufacturing processes; the cost of mining from deeper workings had increased costs; and finally there was a problem in finding suitable labour. The site was sold in 1972 for use as a scrap yard; a sad end for a fascinating industry.

To conclude, we can record a quite different and fascinating use for Bideford Black, developed by the celebrated artist Paul Lewin. He has collecting the raw material and processed it into paints and pastels, then returned to the landscape to create pictures using the pigment.

3.17. The Bideford Black works at some date between 1945 and 1948.
Photo from North Devon Museum Trust Collection.

Chapter 4
THE 'SCYTHESTONE HILLS': The Whetstone Mines of the Blackdown Hills, East Devon

INTRODUCTION

We continue our exploration of the industrial history and geology of Devon's 'lost' mines by journeying to the charming Blackdown Hills, which straddle the Devon-Somerset border south of Taunton (see the map, 1.1, in Chapter 1 for the location of the area). From the culm mines of the Bideford area we have moved geographically about 40 miles to the east, and jumped forward in time about 220 million years. The little-known, almost secret, landscape of the Blackdown Hills is designated an Area of Outstanding Natural Beauty (AONB) (photo 4.1). The Hills form a high plateau, rising to nearly 1000 feet above sea level in the north and sloping gently

southwards. The tableland is deeply cut into by valleys and bounded by steep wooded slopes which pass down into lush valleys and rolling farmland.

In contrast to today's tranquil scene, over 100 years ago the western escarpment of the Blackdown Hills was a hive of activity. At that time the villages of Blackborough and Ponchydown were the centre of a fascinating and little-known industry where miners burrowed into the steep hillsides to bring out masses of sandstone concretions which were fashioned into sharpening stones (whetstones or scythestones). F J Snell, writing in 1904, called these hills the *Scythestone Hills* after the industry for which they were famed.

4.1. *A characteristic view of the western escarpment of the Blackdown Hills, photographed in June 2010. The wooded ridge is formed by the Cretaceous Upper Greensand, and the buildings spread out along its base are part of the village of Blackborough. The old whetstone workings lie hidden in the woodland. The farmland in the foreground is underlain by red mudstones of Triassic age (the Mercia Mudstone).*

Since early times it has been vital to have a way of sharpening metal tools and weapons, and this was provided, until the invention of artificial carborundum 'stones', by the use of suitable abrasive natural stones called whetstones. In pre-industrial agriculture, it was essential to keep scythes and other cutting tools sharp for the efficient harvesting of crops, and it was this mainly agricultural need that gave rise to the local mining industry in the Blackdown Hills that is the subject of this chapter. A farm labourer could get through two or three scythestones a day at harvest time owing to the fact that they broke fairly easily, and once broken were no use for sharpening. It was necessary to sharpen a scythe very frequently, as often as every quarter of an hour, to maintain a keen edge.

The term 'whetstone' is used in a general sense to describe any sharpening or polishing stone which is held in the hand. Scythestones and honestones are both varieties of whetstone. 'A hone is a stone of smooth fine texture used for giving a very fine edge..... A 'scythestone' is a stone of much more closely-defined shape – a thick stick about a foot (30cm) long, probably tapering towards each end – and of considerably coarser texture' (this quotation is taken from a 1983 account of the Ayrshire honestones by D Gordon Tucker). In this chapter I will use the general term 'whetstone' for the stones dug out from the Blackdown Hills, but the more specific term 'scythestone' is found in some accounts. The term 'Devonshire Batts' is also used by some writers, for example, by Robin Stanes in his 1993 account of the industry. The first use of the word 'whetstone', recorded in the Oxford English Dictionary, was in about 725 AD.

The air along the western slopes of the Blackdown Hills would have resounded to the tap of hammers fashioning the stones, together with the sound of sand and rock being tipped down the hill slopes from wheelbarrows bringing the waste out of the mine tunnels. The build-up of many years of accumulated waste from the mines produced an almost continuous belt, formerly visible from many miles away as a distinctive white horizontal stripe along the hillside (it is partly visible on the right-hand side of photo **4.2**). Many 18th and 19th century travellers commented on this feature and were attracted to the workings, which became a notable sight and Victorian tourist attraction. The more intrepid visitors were guided into the underground workings, but a particular attraction was the opportunity to buy some of the beautifully preserved fossils found by the miners as a by-product of the workings (drawings of

4.2. This photo of Blackborough was taken in about 1900. It shows the waste material thrown out from the whetstone tunnels forming a platform, visible along the hillside on the right-hand side, with evidence of tipped material spilling down the scarp face. The spire of Blackborough church (now demolished) can be seen in the distance, and left of it is Blackborough Beacon (once known as Mardle Pen). Photo courtesy of P Planel.

some examples are shown in **4.3**, and photo **4.4** shows fossils preserved in a block of sandstone, of the sort that may very occasionally be found lying about in the area of the whetstone workings). The sale of fossils must have been a very welcome addition to the income of the mining families.

Until about 50 or 60 years ago, the plateau tops and scarps of the greensand hills had a much more open appearance, and the managed plantations that are so characteristic of them today were largely absent. For example, the 1889 1:2500-scale Ordnance Survey map of the Blackborough Common area shows only a very few isolated conifers on the plateau top with the rest of the area covered by rough grassland and bracken. A photo of about 1900 (**4.2**) also seems to show unwooded slopes, with trains of waste spilling down them. Perhaps the lack of woodland was partly related to the fact that any

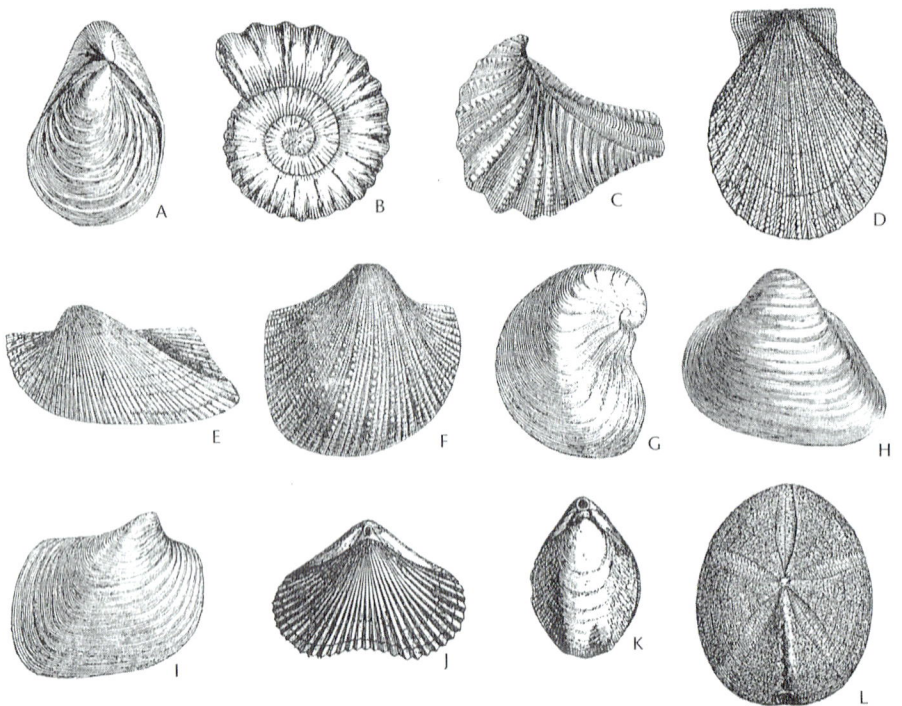

Left: *4.3. Drawings of a selection of fossils found in the Upper Greensand. Not to scale.*
A: Actinoceramus concentricus *(bivalve).*
B: Hysteroceras varicosum *(ammonite).*
C: Pterotrigonia aliformis *(bivalve).*
D: Spondylus dutempleanus *(bivalve).*
E: Nannonavis carinata *(bivalve). F:* Cardium (Granocardium) proboscideum *(bivalve).*
G: `Exogyra conica' *(bivalve). H:* Cucullaea glabra *(bivalve). I:* Panopaea mandibula *(bivalve). J:* Rhynchonella dimidiate *(brachiopod). K:* Ovatathyris ovata *(brachiopod).*
L: Catopygus columbarius *(echinoid).*

Below: *4.4. Fossils preserved in a piece of sandstone from the Upper Greensand of the Blackborough area. The corkscrew-shaped fossils are gastropods. This sample came from the January 1993 excavations mentioned on pages 61 and 69, possibly from Bed 8 or 9 of Downes (see page 66). The original calcareous shell material has been replaced by silica. From top to bottom of the photo represents about 2 inches on the original specimen.* Photographed from a specimen in the collection of Allhallows Museum, Honiton.

trees of suitable size growing nearby would soon have been felled for the timbers which were necessary to support the whetstone tunnels.

Former whetstone workings are present along the entire scarp of the Blackdown Hills between Bodmiscombe in the north and Hembury Fort in the south, and are concentrated especially between Blackborough and Hembercombe (see the map, 4.8). They are marked on some Ordnance Survey maps as 'Levels'. Details of individual sites are given in the Sites and Monuments Record held at the Environment Department at County Hall, Exeter. The main physical feature of the surviving whetstone mining landscape is an extensive platform that extends along and beneath much of the scarp face, and which consists of waste material wheeled out from the whetstone tunnels in barrows and tipped over the slope below. Initially, the waste fan from each tunnel would have been separate, but as working proceeded over the course of many years, the waste fans from adjacent tunnels coalesced to form a continuous apron of debris. There is no detailed topographical survey of the spoil platform, but in the area of Rhododendron Wood [ST 095 072] along the western side of the escarpment, about a mile south of Blackborough, I estimated that the level top of the platform is between 45 and 75ft wide (locally 90ft). No doubt there will be considerable variation in the dimensions of the spoil platform along the whole length of the worked outcrop. In the case of the workings around Hembury Fort, which was probably the latest area to be worked, there are individual linear areas of waste which have not joined to form a continuous platform.

The entrances to the whetstone tunnels are no longer open and accessible, but the sites of the openings can be identified in some places as very indistinct hollows on the hillside. Rotting or removal of the timber supports after the tunnels were abandoned would lead to collapse of the entrances which would then be further concealed by downwash from the hillside above. An attempt in January 1993 to use a mechanical excavator to open a whetstone tunnel mouth, on land on the north side of Blackborough Common [ST 0998 0947], proved unsuccessful since as soon as material had been removed by the excavator, the sands above immediately collapsed onto the cleared area (there is more about this excavation on page **69**).

The other notable feature of the remaining whetstone mining landscape, emphasised by Richard Tamplin in his 2002 study, is the presence along parts of the surface of the spoil platform of hollows and ridges that extend out at right angles to the scarp face (photo 4.5). These are fairly clear on vertical aerial photographs taken by the Royal Air Force in 1947 which are of particular value since they record a time when there was little or no woodland on the plateau top or scarp face, and consequently the remains of the whetstone industry are quite clearly visible. Each linear hollow on the spoil platform represents the remains of a 'barrow run' which extends out from the mouth of a whetstone tunnel and along which waste was barrowed out and tipped out over the edge of the platform onto the slope below. The continuous passage of loaded barrows would have worn a hollow, each of which would be bounded by parallel ridges of waste material. In Rhododendron Wood, mentioned above, I estimated that the tunnels (indicated by the position of individual barrow runs) were spaced at 12 to 15ft intervals, but doubtless the spacing of tunnels varied along the length of the whetstone outcrop.

Richard Tamplin carried out a levelling survey of a sample area [ST 098 094] on the northern scarp of Blackborough Common, to

4.5. Hollow and ridge structures near Blackborough, probably formed by barrow runs extending out from the former openings to whetstone tunnels.

establish the size of the distinct parallel hollows and ridges there. A cross section across the ridges showed them to be between 2ft 4in and 7ft 6in higher than the intervening hollows (the average difference is 4ft 7in), each of which was the approach to a mine entrance. The axes of the hollows were about 17ft apart and this would consequently also be the spacing between individual tunnel entrances (slightly greater than that on the western side of the escarpment at Rhododendron Wood, noted above). The ridges were made up of small fragments of stone in a matrix of soft sand, and the large waste tips, although rarely seen in section, are likely to be made up of similar material. At the same locality, Richard Tamplin found that the waste tip on the hillside below the ridge and hollow structures was about 66ft high from top to toe and extended 39 to 43ft out from the hillside at the top.

The white hillside spoil platforms can no longer be seen from a distance, being now hidden beneath conifer plantations, rhododendron thicket and other vegetation, but they survive as shelves along the hillside along which it is possible to take a variety of pleasant walks (photo 4.6). There is no clear evidence on the plateau surface of subsidence related to the collapse of the tunnels after removal of the timber supports. Any subsidence in the Blackborough area would have to propagate through the 50 to 60 feet of strata overlying the tunnels before being evident at the surface. The amount of any subsidence at the surface is difficult to determine in the absence of detailed knowledge of the layout of the tunnels beneath the hill, but if all the tunnels are the 'standard' dimensions of about 4ft by 6ft, the amount of theoretical subsidence

is unlikely to be great. Robin Stanes reported workings at Hembury and at Newcombe (see the map, 4.8, on page 64) which 'show signs of very recent collapses but this may be due to the gentleness of the slope here and the consequent initial shallow depth of the tunnels'.

It is generally thought that the folly of Garnsey's Tower, on the Newcombe side of Blackborough Common, has been affected by subsidence, and an 1854 watercolour sketch by P O Hutchinson showed the tower cracked from top to bottom (see 4.7). Another explanation may be simply that it was badly built in the first place, with inadequate foundations.

Garnsey's Tower, Blackborough Hill. — Coloured there Sep. 12, 1854.
There are the remains of two floors inside this tower, a staircase and two fire places. The walls are two feet thick, the diameter of the building 12 feet; but so cracked and ruinous, as to threaten great risk of falling.

4.7. *A watercolour sketch of Garnsey's Tower, near Blackborough, made by P O Hutchinson on 12 September 1854. It is possible that the cracks in the tower were caused by subsidence of whetstone tunnels, or it could simply be that the foundations were poorly constructed.* Reproduced by courtesy of the Westcountry Studies Library, reference Z19/2/8E, page 29.

Left: 4.6. *This photo, taken in June, shows a typical path in Rhododendron Wood (ST 095 072), along one of the old 'sandbeds' formed by material thrown out from the whetstone tunnels.*

F J Snell reported a story that Blackborough Beacon had subsided by as much as 30ft owing to undermining by whetstone tunnels, but subsidence of this order seems most unlikely. Another more probable explanation is that Blackborough Beacon has been moved downwards by a fault between it and the main mass of Blackborough Common. The base of the greensand at Blackborough Beacon lies at a height of about 740ft while the base of the greensand in the adjacent Blackborough Common mass is about 50ft higher, at about 790ft. This suggests that Blackborough Beacon is faulted down by about 50ft relative to the main mass of greensand.

In a fascinating account of a visit to the whetstone mines in 1866 the Revd Kirwan writes that the white stripes of waste material formed along the hillside seem to have been regarded locally as a way of forecasting the weather and 'serve the part of a barometer to the entire neighbourhood. When the belt is white fine weather may be expected; on the contrary, as the hue of the belt approaches towards a brown tint, so is the finger of the barometer supposed to point more persistently to rain'. The reason for the colour change is that the sand absorbs moisture when the air is damp and dries out to a paler colour when the air is drier. Whether there really is a forecasting element to this or whether the darker colour of the sand merely represents wetting due to rain which has already arrived, is uncertain – now that the white stripe is overgrown with vegetation there is no way of testing the idea. Another local weather-related belief, recorded by Miss L C E Chalk, daughter of the Revd E S Chalk, a long-serving rector of Blackborough, was that 'It was a sure sign of rain when the donkeys all came to the edge of the hill and brayed'!

The whetstone industry of the Haldon Hills

Most accounts of Devon whetstone mining concentrate, quite naturally owing to its importance, on the industry of the Blackdown Hills. However there is some evidence from the literature that there were also whetstone workings in the Haldon Hills, about 5 miles southwest of Exeter (see the map, **1.1**, on page **10**). They started at some date before 1797 and are recorded as over by some date before 1839. Richard Polwhele wrote in his 1797 *History of Devonshire* that: '....whetstones, of various excellence, are found on the eastern declining sides of Haldon and in some of the chasms upon the down ..... Sometimes the stones are found large enough to form five or six [whetstones] which after being brought to the required form, are carried to a ready sale at Exeter. And those of the best quality are sold at so high a rate as sixpence each. The whetstones that are found on Haldon are generally met with at the depth of three fathom'. The Lysons, in their *Magna Britannia* of 1822, also noted whetstones occurring on the east side of Haldon 'in the parish of Kenne'. Henry De la Beche, writing in 1839, referred to the Haldon whetstones but added that '....we believe that the preparation of them in that locality is now discontinued'. This forgotten industry is not precisely located by Richard Polwhele, except as being in the parish of Kenn. If the industry had been long-lasting and of significance, it would be expected that benches of waste like those so characteristic of the Blackdown industry would have been produced along the hillsides, but there is no obvious record of such features on Haldon, nor of any ridge and hollow structures or other clear evidence of old workings.

DEVON IN THE LATE CRETACEOUS
100 MILLION YEARS AGO

We have seen in Chapter 1 how geologists came to recognize that the surface of the earth is made up of gigantic 'plates' which are in constant, almost imperceptible, motion relative to each other. These movements produce continuous changes in the distribution of the land on the earth's surface through the course of deep geological time. About 240 million years ago, the landmasses of the world were lumped together into a giant 'supercontinent' called 'Pangaea' (meaning 'all lands'). However, about 200 million years ago, as a consequence of the movement of the great 'plates' described above, Pangaea began to break up. By the time of the formation of the concretions that were worked for whetstones, about 100 million years ago in the Cretaceous period, the arrangement of the continents on the face of the globe was beginning to approach a form close to that of today, but there were still many differences. For example, a large ocean lay between 'Europe' and 'Africa', and 'Europe' and 'North America' were much closer, since the sea floor spreading that gave rise to the Atlantic Ocean had only just begun. At the time of whetstone formation, Blackborough lay at a latitude of about 45° N, similar to that of southern France today, and the climate was warm and humid.

The geology of the Blackdown Hills

During the Cretaceous, world-wide sea levels gradually rose until by the later Cretaceous great oceans covered large parts of the world. About 100 million years ago, the site of what are now the

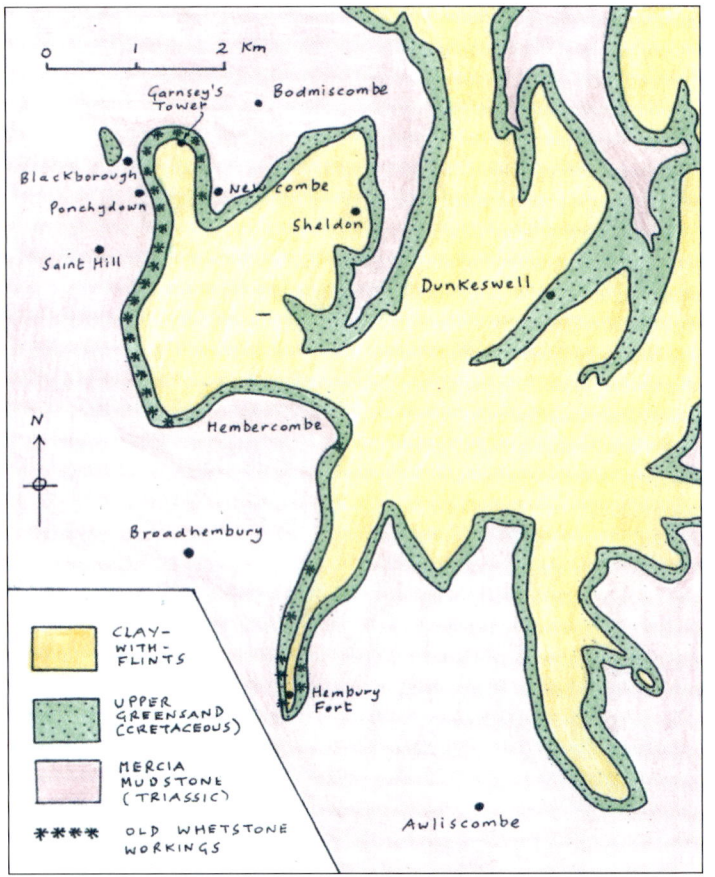

CLAY-WITH-FLINTS

UPPER GREENSAND (CRETACEOUS)

MERCIA MUDSTONE (TRIASSIC)

**** OLD WHETSTONE WORKINGS

Blackdown Hills was covered by the waters of a sea that extended eastwards across the whole of southern Britain into Europe and beyond. On its bed were laid down the sands which now form the Upper Greensand. To the southwest, however, lay an island formed by the upstanding resistant mass of the Dartmoor Granite, for the shoreline of the sea lay only about 10 miles southwest of what are now the Blackdown Hills. What lies beneath the Blackdown Hills and gives them their shape today? Their geology is shown on the map, 4.8, and an imaginary slice (4.9) through the western side of the Hills reveals the geological building blocks that make up their structure. Such a slice, which I have drawn in illustration 4.9, shows the layer-cake structure of the hills. There are three main layers:

Clay-with-flints. The topmost layer, or 'icing' to the cake, covers the plateau top and is a relatively thin skin (mostly between 3 and 16ft thick), of reddish brown clay with unworn flints, called the 'Clay-with-flints'. The unworn flints were left behind after the erosion and solution of the Chalk, and prove the remarkable fact that a layer of Chalk originally covered the whole of southwest England. The only remains of this thick Chalk layer are the resistant flints left behind.

Left: *4.8. This sketch-map shows the geology of the western Blackdown Hills and the location of the main whetstone workings.*

Below: *4.9. A 'slice' through the western Blackdown Hills to illustrate their geological structure.*

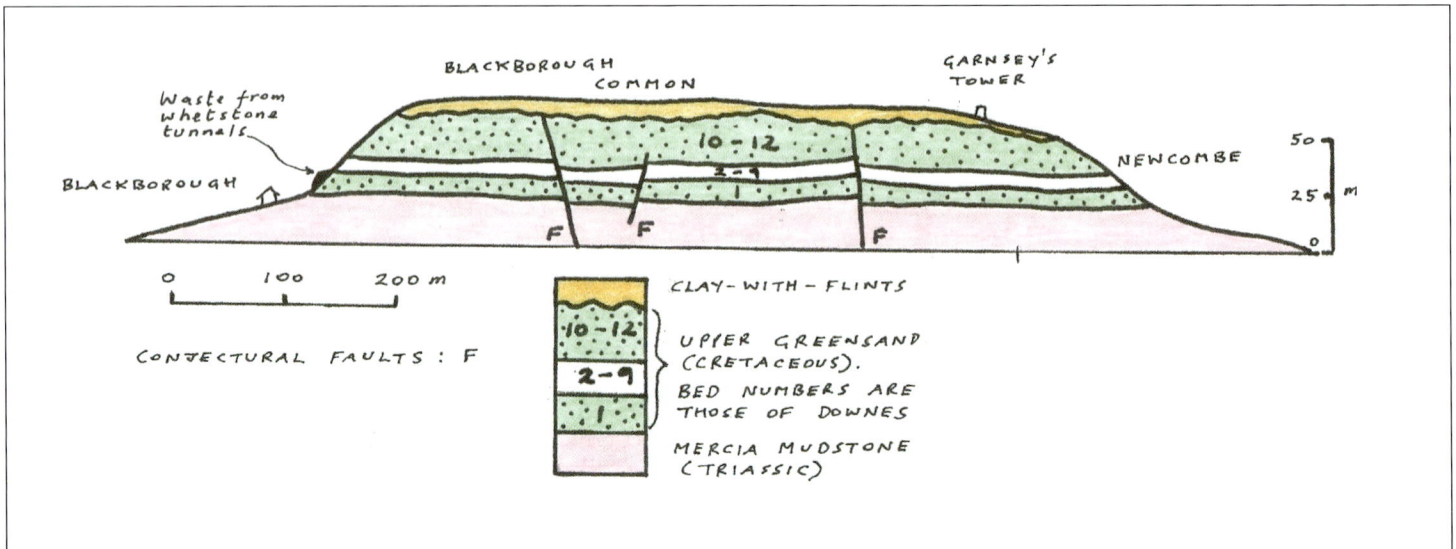

CONJECTURAL FAULTS : F

CLAY-WITH-FLINTS

UPPER GREENSAND (CRETACEOUS).
BED NUMBERS ARE THOSE OF DOWNES

MERCIA MUDSTONE (TRIASSIC)

Upper Greensand. The main part of the geological 'cake' is of most interest to us, since it is from this layer that the whetstone concretions were mined. It consists of about 100ft of sands belonging to the deposit called the Upper Greensand, mentioned above, which we will look at in a little more detail later. The sands give rise to the slopes that form the steep western and northern escarpments of the Blackdown Hills. The sands are water-bearing and may be likened to a sponge; the water travels down through the sands until it meets the lowest level of the layer-cake, which is red mudstone of the Triassic Mercia Mudstone (see below). The water cannot easily penetrate into the mudstone layer and is thus thrown out as springs at the base of the greensand layer.

Mercia Mudstone. The Upper Greensand rests on the basal and very different part of the cake – the Mercia Mudstone. This layer consists of red clays and a more cemented type of clay called mudstone. The Mercia Mudstone belongs to the Triassic period of earth history and is about 250-200 million years old, much older than the Cretaceous Upper Greensand. There is thus a great time gap between the Upper Greensand and the Mercia Mudstone and in this gap we are missing the whole of the Jurassic period, representing over 100 million years of earth history. How did this come about? The explanation is that, after the Triassic, Jurassic and early Cretaceous rocks were deposited as flat layers, they were tilted by earth movements, and vast amounts of the younger rocks were worn away. In the case of the Blackborough area, all the early Cretaceous rocks, the whole of the Jurassic rocks, and the upper part of the Triassic Mercia Mudstone were eroded. Then, about 100 million years ago, a global rise in sea level caused the sea to advance onto this area to deposit the sands of the Upper Greensand. The buried surface between the older and the younger rocks is called an unconformity.

The Upper Greensand

Distinctive units of rocks with broadly unifying features are called 'formations' by geologists and they are the fundamental units shown on geological maps. Many of them can be further divided into smaller distinctive subunits called 'members'. The Upper Greensand is an example of a formation, which extends from Devon across southern England into continental Europe. In Devon it has been divided into three members, consisting, from bottom to top, of the *Foxmould*, *Whitecliff Chert*, and *Bindon Sandstone*. The names of

formations and members are derived either from the localities where they are best developed or were first described (the 'type locality') or from some other characteristic of the unit, such as the main type of rock that it is made up of. The Foxmould Member has its type locality at Whitecliff, near the charming seaside village of Beer in east Devon. The origin of the name is uncertain, but it either derives from the yellowish-brown ('foxy') colour of the weathered sand, or from the fact that earths occupied by foxes are not uncommon along the outcrop of the member. The Whitecliff Chert Member is named from the type locality, also at Whitecliff near Beer, and contains layers of tabular and nodular cherts which are siliceous, rather like flints. The Bindon Sandstone Member is named from the type locality at Bindon Cliffs, east of Seaton.

West of a line between Sidmouth and Yarcombe, in the area which includes the Blackdown Hills, the Upper Greensand is different from the rest of the formation farther east and is informally called the 'Blackdown Greensand'. The calcareous content seems to have been mostly leached away, probably because the protective capping of Chalk which is present farther east has been stripped away by erosion. Another difference is that the Whitecliff Chert and Bindon Sandstone are missing from the Blackdown Hills, where it appears that only the Foxmould is preserved, although in places a few feet of chert beds (Whitecliff Chert) can be seen at the head of some deeply incised valleys, or 'goyles'.

We can briefly summarise the main characteristics of each member. *The Foxmould* in areas outside the Blackdown Hills is 80 to 100ft thick and consists of weakly cemented sandstones which are greenish-grey to bright green when fresh, but commonly weather to a yellowish-brown colour. They contain small rounded green grains of the mineral glauconite, usually considered to have formed on the sea bed. The sandstones have a calcareous cement, but westwards towards the Blackdown Hills that disappears and in the Foxmould of the Blackdown Hills there is very little carbonate – fossils which were originally calcareous have mostly been replaced by silica. And of course the Foxmould of the Blackdown Hills contains the celebrated whetstones. The *Whitecliff Chert*, if present at all, is seen in the Blackdown Hills as only a few feet at the top of the local succession, just below the cover of Clay-with-flints. Elsewhere in Devon, it is 40 to 60ft thick and consists of calcareous sandstones with many layers of nodular or tabular cherts. The cherts have been widely used as a building stone in east Devon. The *Bindon Sandstone* is 10 to 26ft thick and consists of calcareous sandstones. It is especially notable as the source of the building stone ('Salcombe Stone') used in Exeter Cathedral.

Dividing up the Upper Greensand

The first important geological account of the Blackdown Greensand was that of William Fitton in 1836. He gave a detailed description of the strata (between 12ft 6in and 18ft thick) worked for whetstones, including the local miners' names for the individual beds (these are given in **bold** type in the table, 4.10). In 1882 William Downes produced a fairly detailed description of the beds for the whole of the Upper Greensand of Blackdown, and this is shown in 4.11. He numbered the beds from 1 at the base to 12 at the top, and showed the equivalence of his numbered beds with those of William Fitton. The details are shown in the table. The fossil names used by William Downes have in many cases been superseded by more modern names, but I have not changed them.

Fossils and life in the Upper Greensand seas

As we have seen, the Blackdown Hills are famous for the beautifully preserved fossils that have been collected from the Upper Greensand, and many of these were purchased from the whetstone miners by Victorian geologists and collectors, such as William Downes and William Vicary. The original calcareous material of the shells has generally been replaced by hard silica (the process of 'silicification') which has preserved beautiful details of ornament. There are important collections in the West Country museums of Exeter (Downes collection), Bristol (Miller collection), and Taunton (Fox and Williams collection). Other significant collections are in the Sedgwick Museum, Cambridge (Meyer and Wiltshire collection), Natural History Museum (Vicary collection) and British Geological Survey (Sclater collection). William Fitton gave a list of fossils, as did William Downes in 1882.

Top right: *4.10. Dr William Fitton's 1836 subdivision of the whetstone beds.*

Bottom right: *4.11. The subdivisions of the Blackdown Greensand recognised by William Downes in 1882. The beds are numbered from 1 at the bottom to 12 at the top. The equivalence of the numbered beds with those given by William Fitton in 1836 is shown.*

1. Reddish sand rock, extending upwards to the top of the hill. [Thickness not given].

2. **Fine vein**. Concretions, the best for whetstones. Very fossiliferous. 2in to 1ft.

3. **Top sand rock**. Sand with irregular concretions, no use for whetstones. 3ft to 4ft.

4. **Gutters**. Concretions in 4 or 5 layers in sand. The bed most commonly worked for whetstones. 3ft to 5ft.

5. **Burrows**. Stone and sand, used only for building. 2ft to 3ft.

6. **Bottom stone**. A range of concretions, excellent for whetstones. 2in to 6in.

7. **Rock sand**. Sand with fewer concretions; no use. 4ft.

8. **Soft vein**. Concretions excellent for whetstones. 2in to 6in. Strata below 'not known to the workmen'.

Bed 12. Sand, with layers of cherty sandstone passing upwards into chert. Characteristic fossil *Pecten quadricostatus*. About 25ft.

Bed 11. Fine sand, variegated, with thin lenticular and impersistent shell bands about 2in thick. *Petunculus sublaevis* and *Trigonia affinis* abundant. About 18ft.

Bed 10. Red rock [equals Bed 1 of Fitton, in part]. Sandstone in layers divided by sand. Very fossiliferous, with 46 species recorded. *Cyprina cuneata*, *Exogyra conica* and *Cucullaea carinata* the commonest fossils. About 3ft.

Bed 9. Hard fine vein [equals Bed 2, Fine vein, of Fitton]. A thin layer of concretions used for scythestones. (Curiously, although Fitton records abundant fossils in this bed, they are not mentioned by Downes). 2in to 1ft.

Bed 8. 'Clumps'. Sand, very fossiliferous. The predominant fossil is *Turritella granulata* in 'clumps'. About 2ft.

Bed 7. 'Clumps'. Sand, very fossiliferous. The predominant fossil is *Petunculus umbonatus* in 'clumps'. About 2ft. [Bed 8 and Bed 7 together equal Bed 3, Top sand rock, of Fitton].

Bed 6. Gutters [equals Bed 4, Gutters, of Fitton]. Sand and concretionary layers. Few fossils, including *Inoceramus sulcatus* and *Petunculus umbonatus*. 3ft to 5ft.

Bed 5. Burrows [equals Bed 5, Burrows, of Fitton]. Concretions divided by sand layers; mostly used for building, sometimes for whetstones. About 4ft.

Bed 4. Bottom stones [equals Bed 6, Bottom stone, of Fitton]. Concretions used for whetstones. Up to 5ft.

Bed 3. Bottom rock [equals Bed 7, Rock sand, of Fitton]. Sand with very few fossils, including *Trigonia aliformis*. About 4ft.

Bed 2. Soft fine vein [equals Bed 8, Soft vein, of Fitton]. A thin layer of concretions, used for scythestones. Up to 6in.

Bed 1. Whitish brown sand rock, unfossiliferous. The 'white rock' of the miners. About 30ft.

Most of the fossils found are gastropods (sea snails) and bivalves (such as oysters and clams). At least 65 species of gastropods and 83 species of bivalves were recorded by J D Taylor and his colleagues, writing in 1983. Many of the fossil bivalves and gastropods show holes where predatory gastropods have drilled into their shells. Ammonites, often beautifully preserved, have also been found in the Blackdown Greensand. They are the most useful fossils in this part of the geological column, as well as the Jurassic, for dividing it up the strata into what geologists call Stages, which are further subdivided into Zones and Subzones. William Downes also listed other fossils, including a few brachiopods, echinoderms (sea urchins) and annelid worms (mostly serpulids). Corals, common in the shallower-water Upper Greensand of the Haldon Hills, are barely represented in the Blackdown Greensand. Sponges littered the sea floor, and as we shall see below, probably played an important part in the formation of the whetstone concretions. William Downes noted that sponge spicules are common throughout the Blackdown Greensand but were especially common in Bed 5 ('Burrows') (table 4.11).

The gastropods and bivalves lived on a sea bed at water depths of between 50 and 280ft, the bivalves mostly burrowing to shallow depths into the sand and mud of the sea floor. J D Taylor and others have made a more precise comparison with a modern community, living on a sand bottom at water depths of 130 to 280ft off the Atlantic coast of Spain, which seems to show many similarities to the ancient faunas of the Blackdown Greensand. The gastropods and bivalves are mainly found in 'clumps', and the two valves of bivalves are commonly still joined together; these features suggest that the sea-bed was quiet and mostly undisturbed by waves or currents. However, above the level of Bed 8 ('Clumps') of William Downes (see the table, 4.11), the shells become progressively more worn and broken and in many cases the valves are separated. It seems that above this level the bottom conditions became more turbulent, so that the shells were thrown about and broken. There is some evidence of weak tidal currents and the occasional layers with broken-up shells suggest the influence of rare storms affecting the sea bottom. The sediment was thoroughly churned up by the activities of burrowing organisms, a process called 'bioturbation'.

How the whetstone concretions formed

The precise way in which the whetstones formed is not known for certain. Sponges were common on the Cretaceous sea bed in the Blackborough area, and on their demise, their soft parts decayed and the spicules (tiny siliceous rods that stiffen the tissues of sponges) became detached and were scattered over the sea bed. Possibly, local concentrations of spicules became the site of concretions cemented by calcareous material. They may have been selectively cemented because of the original porosity imparted by the presence of abundant spicules. After removal of the overlying Chalk cover by erosion, the carbonate was dissolved, and some of the silica was redistributed as cement to produce the concretions in their present form.

THE HISTORY OF WHETSTONE MINING

Local people must have been aware from the earliest times of the sharpening properties of some of the stones that could be found lying on and around the greensand escarpment, or exposed in the sides of the steep-sided valleys (locally called 'goyles') cut into the greensand. However, the date when someone decided to burrow into the hillside to seek a larger supply of concretions which could be fashioned into whetstones will always remain uncertain. At least one example of a whetstone was found in a Middle Bronze Age enclosure ditch during excavations for the new Honiton to Exeter A30 road (recorded by Fitzpatrick and others in 1999). In a survey of English honestones by D T Moore, only one Blackdown whetstone was recorded, from a 16th-century level in Humberside. These early records do not necessarily prove the existence of contemporary underground workings in the Blackdown Hills, for it is possible that usable whetstones may have been found protruding from the sides of goyles, or simply picked up from the ground.

Robin Stanes has shown that the first reference to the industry was in May 1690, in the Bridgwater Port Books. He noted that the industry was recorded in notes for a parish history of Kentisbeare compiled in about 1755 by Dean Milles. There is an interesting early account by Richard Polwhele in his 1797 *History of Devonshire*, in which he refers also to the whetstone industry of the Haldon Hills (page **63**). Even in 1797 the Blackdown industry was mature since Polwhele refers to the miners discovering old workings during the driving of fresh tunnels. Charles Vancouver gave a brief account in his *General view of the agriculture of the county of Devon* in 1808, with interesting observations on the working methods used in the mines (see page **69**).

19th century geologists found the geology of the Blackdown Hills fascinating, their interest stimulated mainly by the beautiful fossils that were emerging from the whetstone workings (see 4.3 and 4.4). One of the earliest geological descriptions was that of the Swiss-born geologist Jean-Andre de Luc who visited the area in 1806 and wrote an account of his visit in 1811. He stayed with General Simcoe at Wolford Lodge, about 5 miles north of Honiton, and the general accompanied him on a tour of the Blackdown Hills which included the whetstone workings. In 'the upper part of Hembercombe' de Luc noted that the whetstones were worked from two different beds, each composed of sand 'partly consolidated in large concretions', separated by a bed of pure sand. The concretions in the two beds noted were very different: those in the upper level were much softer and of no use as whetstones. They were only worked because they had to be removed in order to get at the good whetstone concretions in the lower level. He gave a good description of the methods of working, which we will look at again below.

As part of his classical and wide-ranging 1836 account *Observations on some of the strata between the Chalk and the Oxford Oolite, in the south-east of England*, Dr William Fitton described the 'sithe-stone' workings of the Blackdown Hills, and gave a list of Blackdown fossils. As we have seen (page 66 and 4.10) he was the first person to give a detailed description of the strata worked for whetstones, which included the local miners' names for the individual beds. William Fitton also gave some details of the methods of fashioning the 'sithe-stones', which we will look at in more detail below. Other brief Victorian accounts include those of the Lysons in 1822, Moore in 1829, Shirley Woolmer in 1831, Knight in 1836 and the Sidmouth antiquarian P O Hutchinson in 1854. The short account given in Henry De la Beche's celebrated 1839 memoir on the geology of southwest England seems to have been largely based on the accounts by de Luc and Fitton.

A story published in the American magazine *Harper's New Monthly Magazine* for 1854 is a dramatic moral account set in the Blackborough area, which revolves around a shortage of wooden props for the whetstone tunnels, the sale of them to buy drink, and the death of a miner following the collapse of a tunnel after the props had been removed.

After William Fitton's account in 1836, some considerable time passed before anyone looked again at the geology of the whetstone beds in detail. The next significant account (see page 66) was that of the Revd William Downes who wrote in 1882: 'Living, as I do,

almost at the foot of the Blackdown Hills, it has naturally been my aim and ambition to add something to our knowledge of their much discussed and highly fossiliferous strata'. He went on to emphasise the difficulty of the study, owing to the lack of good sections in the strata, the fact that only a very few pits were still open (in 1880), and the confused way in which the fossils were collected. The miners would bring the fossils home and jumble them all up together, making it very difficult to establish from which particular beds they came. An important point is that all fossils collected by the miners naturally came from only the relatively thin interval of sands (up to 18ft out of the total Upper Greensand thickness of 100ft) worked for whetstones. Nevertheless, in spite of the uncertainties mentioned above, William Downes was able for the first time to produce a fairly detailed description of the beds for the whole of the Upper Greensand of Blackdown, which has not been improved upon since. As we have seen (page 66 and 4.11) he numbered the beds from 1 at the base to 12 at the top, and showed the equivalence of his numbered beds with those of William Fitton. Downes' collection of fossils from the Blackdown Hills is now in the Royal Albert Memorial Museum, Exeter.

In 1894 R D Blackmore, the famous author of *Lorna Doone*, wrote a novel called *Perlycross* which includes some fictional scenes involving whetstone miners. In April 1899, Horace B Woodward led an excursion party of Victorian geologists to the Blackdown Hills. From Cullompton Station they walked to the Ponchydown Inn where they purchased 'a few fossil sponges and some echinoderms [sea urchins]'. They visited a whetstone level northeast of the post office on Blackborough Common, apparently the only working then open. The descriptions by Jukes-Brown and Hill in their account of the Cretaceous rocks of Britain in 1900, and that by Ussher in the 1906 Geological Survey Memoir for the Wellington district, largely repeated the work of William Fitton and William Downes.

In 1904 F J Snell, writing in his *Early Associations of Archbishop Temple*, gave some interesting details of the industry. By Snell's time the industry had declined to the extent that only two mines were still in operation. The Revd E S Chalk noticed the mines in his book and papers on Kentisbeare and Blackborough (1910 and 1934). The Revd Chalk wrote in 1910, when the industry was close to fading out: 'As the industry is almost extinct a short account of it may be of service. Level galleries, some of them two or three hundred yards in length, were driven in the greensand. At the height of the trade there were about twenty-four pits in working, employing two to four men each, besides women. Every inch of the gallery has to be propped up, and the work at the end of the level is very dangerous. The stones were rough hewn at the mouth of the pit, and were driven in wheelbarrows by lads and women to the sheds, where they were finally shaped with a strange tool, like a stout hammer with a double head beaten into blades; they are still made locally. In course of time the limited district of the stones was riddled through and through, and the three pits opened during the last six years cut across the old workings continually.'

Few geologists have studied the Blackdown Greensand in detail since the time of William Downes, but in 1983 J D Taylor and his colleagues looked in detail at the behaviour of the gastropods (sea snails) which preyed on other gastropods and bivalves on the bed of the Blackdown Greensand sea. This study was based on the many beautiful specimens available in museum collections, but care has to be taken in studying such collections since many fossils in museums labelled as 'Blackdown' came from other localities. One way of telling is to take into account that all Blackdown fossils are generally silicified (that is, the original calcareous material has been replaced by silica), so that if a calcareous fossil is labelled as 'Blackdown' it is most unlikely to have come from there.

In January 1993 Mr George Bate of Honiton, an enthusiastic amateur geologist with a keen interest in the whetstone mines, funded an attempt to dig out one of the old tunnels, using a mechanical excavator, on land owned by Ray Northam of Bodmiscombe Farm. It proved impossible to re-open an old tunnel, but the scientific results were of great value (they were described by Mark Woods and Neil Jones in 1999). The strata excavated around the former mine entrance consisted of about 23ft of sands which could be compared with beds 7 to 11 of William Downes (see 4.11). This was the first study of the Blackdown Greensand in situ since the days of Downes. Examples of fossils collected from the excavations are shown in photo 4.4.

The most recent, exhaustive and fascinating account of the whetstone mining industry and the social history of the mining community of Blackborough was produced by Robin Stanes in 1993. The present account owes much to Robin Stanes' work, and I am indebted to him for permission to draw on selected parts of his description of the industry.

THE WHETSTONE MINES

The tunnels from which the whetstones were extracted were dug horizontally by hand using pick and shovel, into the hillside for about 500ft (about 167 yards) according to Jean Andre de Luc, reporting on his 1806 visit, but about 300 yards according to William Fitton and up to 400 yards according to Vancouver. A figure of 200 to 300 yards was given to me in 1984 by Percy Lane of Blackborough, one of the last people to have personal recollections of the whetstone workings, and this is also the figure given by the Revd Chalk in 1910. William Fitton noted that '.....the drifts are commonly pushed to about 300 yards inwards, greater distances not repaying the labour of bringing out the sand'. The spur of Upper Greensand at Blackborough Common is 500 to 730 yards wide. If we accept the tunnel lengths given by Fitton, it is possible that some workings from the western (Blackborough) side might have met with workings driven in from the eastern (Newcombe) side. At Hembury Fort, the greensand ridge is only 330 yards wide and it is likely that workings from either side met beneath the hill.

Charles Vancouver, writing in 1808, has some interesting observations on the methods of working the mines. He wrote: 'The manner of working these quarries is by driving a road or level from the side of the hill to the distance of 3 or 400 yards, about three feet wide, and five feet and a half high; when the hill has been penetrated to this distance the usual practice is to work out all the loose sandstones within eight or ten yards of the road, leaving pillars at first to support the roof of the mine, but afterwards gradually working them out, and suffering the whole excavation to fall in and fill up after them. The size of these stones seldom exceeds that of a horse's head, all of them more or less grooved and indentated...' It is reported that the tunnels were widened and heightened internally, and later widened further and cross-galleries dug to link tunnels to extract more stones. However it is possible to doubt Vancouver's suggestion that pillars were left to support the roof, but were afterwards gradually worked out and the whole excavation allowed to collapse, since it is unlikely that the relatively weak Upper Greensand would be strong enough to stand for long unsupported; it would depend possibly on the size of the pillars.

Only one photograph (photo 4.13), is known showing the inside of one of the tunnels. It is difficult to be precise about the dimensions given the lack of scale to the photo, but Jean de Luc noted that the galleries are 'the height of a man' and 'of a breadth

Mode of trimming Whet-stones for scythes on the Whet-stone Hills, Punchey Down, and one of the entrances into the side of the hill. This passage entered 300 yards horizontally. Sketched Sept.ʳ 12. 1854.

4.12. *An 1854 illustration by Peter Orlando Hutchinson of a miner posing at the entrance to a whetstone tunnel. He is holding a basing axe, and beside him is a pile of whetstones.* Reproduced by courtesy of the Westcountry Studies Library, reference Z19/2/8E, page 31.

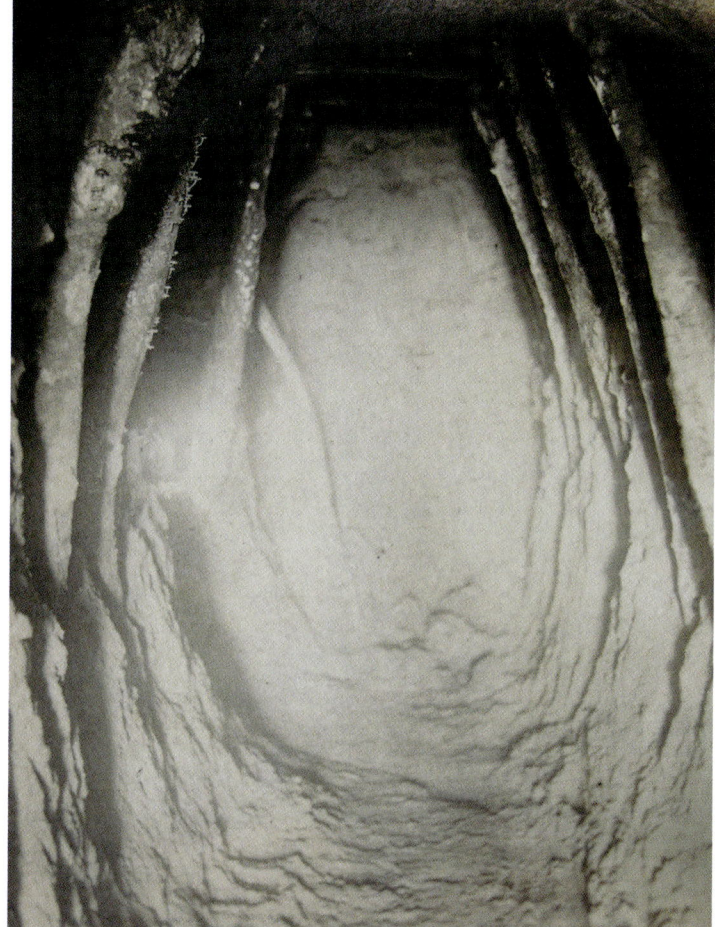

4.13. A photograph, the only one known, of the inside of a whetstone tunnel. Reproduced by permission, Devon Record Office, reference 3269Z/1, page 5.

American magazine of 1854 (*Harper's New Monthly Magazine*) the plot of which hinges on the theft and sale of pit props to buy drink and the death of a young miner as a result of a fall in an unsupported tunnel. After a tunnel was finished, the supports would be pulled out for use in a new tunnel, and the original tunnel would collapse.

A scene of a miner at the entrance to a tunnel, drawn by the celebrated Sidmouth antiquarian Peter Orlando Hutchinson, shows the miner using a basing axe (described below) to work a whetstone nodule (illustrations 4.12 and 4.14). A pile of completed whetstones lie beside the miner, although these were fashioned elsewhere, not at

4.14. A detail from 4.12, showing the tools used by the whetstone miner; in the front is a basing axe, with a type of pick behind it. See also the drawing, 4.17. Reproduced by courtesy of the Westcountry Studies Library, reference Z19/2/8E, page 31.

sufficient to allow two men with wheel-barrows to pass each other'. According to Percy Lane, the wheelbarrows were only about 2ft wide to allow passage in the narrow tunnels. F J Snell wrote, in his 1904 *Early Associations of Archbishop Temple*, that 'The galleries were propped up with posts, and at intervals there were "thirt-holes", where a whetstoner would stop and draw up his barrow, to allow another to pass'.

It can be seen from the photo (4.13) that the strata were supported at intervals by wooden supports, slightly curved inwards, with a horizontal cross-beam at the top. The supply of sufficient amounts of suitable timber for supports must have been a continuous problem. I mentioned earlier the vivid story in an

Above left: *4.15. A miner, believed to be John Rookley, the last whetstoner, sitting at the entrance to a mine holding a basing axe and a whetstone; the date is uncertain, but may be sometime in the 1920s.* Photo courtesy of Derrick Rugg.

Above right: *4.16. A miner at the entrance to a whetstone tunnel. Information from the Tiverton Museum of Mid-Devon Life identifies this miner as a Mr Radford, rather than John Rookley as previously supposed. The date is uncertain, possibly 1920s.* Reproduced by courtesy of the Tiverton Museum of Mid-Devon Life.

the mouth of the mine. In photo 4.15 a miner, possibly John Rookley, is shown seated at the mouth of a tunnel, using a basing axe. Another photo (4.16) also shows a miner, probably a Mr Radford, standing at the entrance to a whetstone tunnel.

There are no plans in existence of the underground workings, and it is indeed very doubtful whether any surveys were ever carried out to record the workings. Mining proceeded in an unplanned piece-meal fashion and consequently there were no records of earlier workings available to the miners. As early as 1797 Polwhele wrote that 'the workmen are now in many places driving fresh pits over those already worked'. In 1806 Jean de Luc also recorded that the miners sometimes met with ancient tunnels in a different direction of which neither the date nor starting points were known.

Little was known about how the miners were given the rights to dig for whetstones until in 2000 Robin Stanes gave an account of nine leases of land for whetstones dating from 1788 to 1800, written into one book and one loose sheet (Devon Record Office 56/10 box 12/23-24). The leases were granted by the Drewe family who lived at The Grange, Broadhembury, and gave the right to individual miners or to a pair of miners to open and work a 'whetstone pit' for a period of, normally, two years. The rent, payable by instalments quarterly, was between 12 and 25 guineas (25 guineas was about the yearly income of a labourer at that time). The terms of the various leases were different, but generally speaking they included the following requirements: the lessee should keep within the boundaries laid down and not encroach on neighbouring workings; the pits should be dug in the usual custom and manner; the common should not be damaged; the miners should not work more than one pit at the same time; should not use more than two wheelbarrows in each pit; should not take dogs to the pits; and should preserve the game and rabbits of the area. The Revd Kirwan noted in 1866 that the miners purchased mining rights for a sum of eight guineas annually.

William Fitton, following a visit in 1825, described the tools in use by the miners: 'The tools employed are a sort of axe or adze with a short handle [numbered 2 in 4.17], called a 'basing-hammer', which is ground to a sharp edge. These are made at the adjacent village of Kentisbere. The other tools are 'picks', without any peculiarity of structure and 'hollowing-shovels' [numbered 1 in

4.17] for digging the masses of stone out of the sand'.

According to Percy Lane, lighting for the miners to work by was provided by a stake driven into the ground with a hole on the top to hold a lighted candle.

Jean de Luc, in his 1806 account, wrote that after completing one tunnel 'The workmen afterwards begin again, from the entrance of the quarry, cutting to nearly the same breadth as before, but now throwing the sand, and such of the concretions as are useless into the space already cleared'. An objection to this observation is that it is difficult to see how the new tunnel could be adequately supported.

The concretions formed only a very small part of the worked material, so that great quantities of waste were produced. This was tipped out over the hillside, and as we have already seen, over the many years of working the result has been to produce a continuous

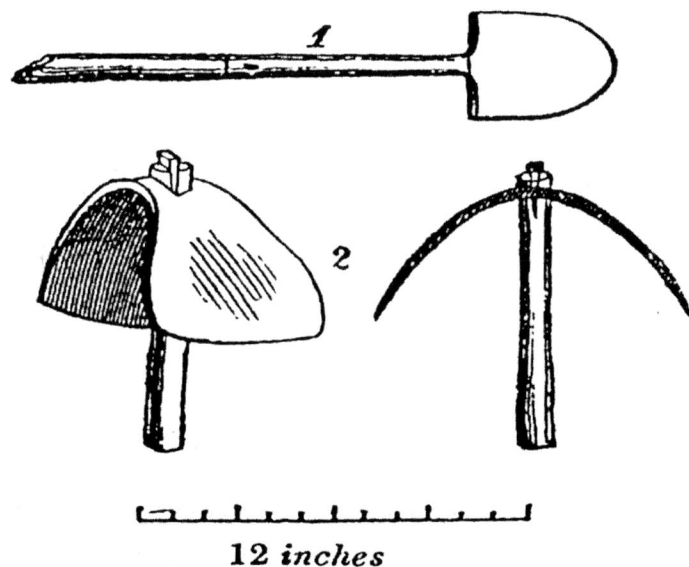

4.17. Tools used by the whetstone miners, reproduced from a drawing in William Fitton's 1836 account.

spoil platform especially along the western escarpment of the hills. Jean de Luc remarked that the workmen built huts on these 'sandbeds' for shelter and to store their tools. The entrances to the tunnels in the later years of the industry had locked doors.

On the Hembury Fort spur the workings appear to have been later than those at Blackborough since they were not mentioned in an 1874 account by Orlando Hutchinson, and no continuous level of waste has developed. Instead, individual elongated mounds of spoil associated with individual tunnels are present, and can be seen clearly on aerial photographs dating from 1947.

In 1806 Jean de Luc followed one of the galleries at Blackborough to the end and interestingly was shown by a miner a place where a fault (a dislocation in the level of strata which throws one side down relative to the other) had affected the whetstone beds, with a vertical displacement of about 5ft. Shirley Woolmer visited the whetstone mines in 1831 and noted that the openings to the tunnels were 5 or 6ft high and 4ft wide, but became wider and higher inside. They extended for 200 to 300 ft, but older ones which had extended for up to 400ft were no longer used owing to the length of time it took to move material by wheelbarrow to the mouth of the tunnel.

An excursion of the Exeter Naturalists' Club to the whetstone mines in 1866 was led by the Revd R Kirwan. The party took the train from Exeter Queen Street (now Central) station to Honiton and from there drove to the mines at Blackborough via Combe Raleigh and Wolford Lodge. They went into one of the tunnels and the Revd Kirwan 'illumined the pit with magnesium wire'. After the visit the party were the guests of Mr E S Drewe, owner of The Grange at Broadhembury, They then drove to Honiton where an excellent tea awaited them at the Dolphin Hotel. The Revd Kirwan gave a talk about the mines, illustrated with specimens, maps and sections. This account is of particular interest because it gives fascinating contemporary information about the detailed working methods and the organisation of the miners. A gang of four men worked together. The leader, No 1, using a short pickaxe, dug out a tunnel 3ft wide by 6ft high, shoring up the gallery every foot of the way with timber props cut from the woods above; 'the looseness of the surrounding material necessitates extreme caution in this part of the work'. The tunnel was dug for a length of 9ft which represented a day's work. The job of No 2 in the gang was to barrow-up the loose stone and wheel it to the end of the tunnel. No 3 wheeled away the waste sand and unsuitable stone and shot it down over the slope. The rate was

4.18. *A rare posed photograph, probably taken between 1860 and 1870, of miners at the entrance to a whetstone tunnel which can just be made out behind the boy with a barrow-load of stones. The man on the left holds a pick and a long-handled Devon shovel, and the man on the right holds a basing axe in his right hand and a finished whetstone in his left hand.* Reproduced by permission, Devon Record Office, 3269Z/1, page 5.

100 barrow-loads a day, the leader keeping a tally in his head. No 4 in the gang checked over the concretions and sorted them: those too small for whetstones were rejected. Other stones were piled up for use as building stone. Suitable concretions were roughly split ready to be shaped into whetstones.

The Revd Kirwan also noted the many faults that affect the whetstone beds, but they seem to have small displacements, generally of about 3ft. He wrote: 'Minor faults everywhere occur: indeed the whetstone veins scarcely ever present a continuous line for an extent of a dozen yards. Here an upcast has occurred, raising the veins to a height of three or more feet above the normal line, and there a down-thrust has occurred, causing a corresponding depression'.

A rare photograph, probably dating from between 1860 and 1870, shows a group of miners at the entrance to a whetstone tunnel (photo **4.18**).

MAKING THE WHETSTONES

Whetstones were fashioned from the concretions found in the Upper Greensand. Because they still had a high moisture content they were relatively soft and could be readily cut or chopped into shape, using a specialised tool called a basing axe or basing hammer. One is shown in use in the illustration by Peter Orlando Hutchinson (4.12 and 4.14), and others can be seen in the photographs of miners (4.15 and 4.18). The man in photo 4.15 may be the last whetstone miner, John Rookley; his basing axe survives in the collections of the Royal Albert Memorial Museum, Exeter. A replica of a basing axe can be seen on the display board near the site of the now demolished church at Blackborough.

William Fitton noted that the concretions were very irregular in shape, but on the outside traces of the original sedimentary layers (bedding) could be seen, and these were useful as a guide to the workers because they were the natural surfaces along which the concretions could be initially split. The concretions varied from 6 to about 18 inches in diameter. Jean de Luc noted that when first taken out they were easily split (along the bedding, as noted) into flags about 2 inches thick, and while still retaining the natural moisture and thus softer and easier to work, were divided into equilateral pieces about 9 inches long.

The method of working was described by William Fitton in 1836. A vertical post was fixed in the ground between the knees of the workmen, with strong pieces of leather to protect the left knee of the workmen. Using a basing axe, long portions were split from the blocks (along the bedding). These were then cut or chopped down on the anvil or knee of the worker 'just as a carpenter cuts timber with an adze'. After this rough shaping, the stones were fashioned to the correct dimensions with a larger hammer. In the same shed would be a group of women whose job was to work the whetstones on their knees by rubbing them into shape against a rubbing stone (an example is pictured on the right-hand side of photo 4.19), constantly dipping them into a tub of water. Jean de Luc wrote that during the process 'an abundance of sparks are continually produced'. When dried they were ready for sale. The size varied from 10 to 12 inches long; some were cylindrical but smaller at each end with a diameter in the middle of about 2 inches; others had the ends and sides cut square. Examples of finished whetstones, and stages in the preparation of the stones, are shown in photo 4.19. Mr Ray Northam of Bodmiscombe Farm has amassed a considerable

collection of whetstones over the years, and he tells me that the finished forms in photo 4.20 are characteristic of those he has collected. However, there was clearly a range of shapes produced, for varieties described as 'long stones', 'short stones', 'long fine stones', 'short fine stones', 'sheg (or shag) stones and 'hill stones' were being sold in the 1780s by Francis Broom, a trader in whetstones from Bodmiscombe. We do not have details of how these whetstone varieties differed in shape, although there seems to have been a basic distinction between long and short forms.

F J Snell, in his *Early Associations of Archbishop Temple* of 1904 wrote that 'To give the raw stones a plain face, women chipped and 'yowed' them with a rubber – *i.e*, another stone – having first placed them in a tub of water to soften them...'

In an interview which I had with Percy Lane in 1984 when he was 82, he gave me a fascinating eye-witness account of the working methods in the 1920s, probably the last person living to be able to do so. He told me that '...they'd take the whetstones, they'd rough chip them [at the mine entrance] they wouldn't want to carry any more than they were forced to, and then they'd take them home and chip

4.19. Examples of whetstones in the collection of Mr Ray Northam of Bodmiscombe Farm. On the right is a rubbing stone on which the whetstones were rubbed with water into their final shape. To the left of it are two stones which have been roughly shaped with a basing axe (pictured in 4.14). Beside them are two examples of stones which have been partly rubbed but have broken in the process and been discarded. Finally, the two stones on the left are completed whetstones (shown in more detail in 4.20). The scale is in inches. Photographed by courtesy of Mr Ray Northam.

4.20. Examples of two finished whetstones, showing front and side views. The scale is in inches. Photographed by courtesy of Mr Ray Northam.

'em a lot more and then their wives used to do the finishing always, they'd have an oak tub up to just where they could reach them, then they'd have a flat big stone leaned up in the tub and then they dipped the stones down in the water and you could hear them rubbing up and down. Somewhere about 8d or 9d, that was the price, it wasn't a lot of money, it was a lot of work'. He said that when he was at school: '....we used to go up there [to the whetstone mines] at dinner times and go in.....if they were gone home from work, well we shouldn't have done I know but we used to go up there in the pits and go in you see sometimes they'd go well 200 or 300 yards you see straight back'.

SELLING THE WHETSTONES

Robin Stanes has estimated that, assuming 24 mines in work with one scythestone cutter for each, it might have been theoretically possible to produce 2,000 whetstones a day, or 10,000 a week, or 500,000 per year. Such production was unlikely to have been

attained because many of the miners worked on the land, especially at harvest times, but even half the theoretical maximum represented a very substantial total. William Fitton, writing in 1836, noted that a good workmen could produce 'several dozen' whetstones each day.

There was a considerable national market for the whetstones, which were also sold locally and exported. In the Devonshire volume of *Magna Britannia* (1822) the Lysons noted that 'The greater part of the whetstones, which are sold under the name of Devonshire batts, are sent to Bridgewater, and thence by water to Bristol, Gloucester, Worcester, &c. &c. Some are exported from Topsham to London'. Shirley Woolmer, writing in 1831, noted that the whetstones were considered the best of the kind in England. Robin Stanes wrote that two whetstone dealers based in Honiton, including a William Munk, were noted in *Pigott's Commercial Directories* between 1823 and 1844. Munk was probably the person referred to by William Fitton when he wrote that the whetstones were sold 'chiefly to one merchant in Honiton who supplied the retail dealers'. Henry De la Beche, writing in his famous report of 1839, noted that 'A large portion of the southern part of England is supplied with these *Devonshire batts*, the sithe-stones of the north being chiefly obtained from Derbyshire and Yorkshire, where considerable numbers are manufactured'.

Robin Stanes has examined the marketing of the whetstones in detail. He noted that the first reference to their export was in the Bridgwater Port Books where in May 1690 the export of 200 dozen scythestones to Bristol is recorded, and these are most likely to have come from Blackborough. In the case of Exeter, he found records of exports from February 1728. Exports continued from these dates in almost every year from both ports until the Port Books end in the 1780s. By 1784 Francis Broom, the whetstone dealer from Bodmiscombe, was buying whetstones and selling them to at least six dealers in London and Southampton. The whetstones were shipped in large quantities from Topsham to London, Portsmouth and Southampton, and smaller quantities to Bristol. Broom seems to have bought 9,700 dozen (116,400) whetstones between 1784 and 1789. His (probably incomplete) account books record the shipment of 3,156 dozen (37,872) whetstones in the period between 1784 and 1788. Robin Stanes gives more details of these shipments in his account of the industry. In 1839 Customs books record the export of 27,731 whetstones from England to Europe, North America, South Africa, Australia, Brazil and the West Indies, but it is impossible to say how many of these came from the Blackdown Hills since no source is shown. It is fascinating to think that in so many parts of the world, just beneath the soil in some field, lie the remnants of a Blackdown whetstone which has been discarded after breaking.

The local market was also important. The 'Scythestone Fair' was an important date in the calendar of the Blackdown miners. It was held in Waterbeer Street, Exeter, once a year in May or June, just before the hay harvest. The sale began very early, at 5 or 6 o'clock in the morning and was all over by breakfast; this gave rise to the old Exeter saying 'all over in ten minutes like Scythestone Fair'. The Revd Chalk has described the scene. The miners were said to have travelled by night driving asses and ponies with panniers laden with scythestones, or else in carts. The distance from Blackborough to Exeter is about 20 miles so that at a speed of say 4 miles per hour the journey would have taken at least 5 hours, meaning a start at around midnight, or earlier. After the sale, Chalk says that the

panniers were loaded with groceries and when they got back to the foot of the Blackdown ridge after their long journey home the miners filled what space was left in the panniers with earth to improve their gardens. An important name associated with the Scythestone Fair was that of a Mr Munk, mentioned above, who had moved in 1844 from Honiton to Exeter where he had an ironmongers shop in Waterbeer Street. In 1844 he was described as a scythestone manufacturer and ironmongers' wholesaler. Strangely, Robin Stanes could find no contemporary description of the Scythestone Fair, which must have been a striking spectacle, except for a note in the Blackborough School log book for 1878 which recorded the total absence of pupils form the school on 14 June – presumably because they were all at the Fair!

In 1866 the Revd Richard Kirwan noted that 'When finished [the whetstones] are sent to the neighbouring town of Cullompton, and command a ready sale. At the present time the demand for these whetstones exceeds the supply: the pits yield much fewer stones than formerly, and apparently the veins are nearly exhausted. The finest sorts of stone fetch about 12s. a dozen; however, I am informed that the pits yield perhaps five hundred dozen of inferior stones for one dozen of the best. Inferior sorts are sold at prices ranging from five shillings to as low as one shilling per dozen.' The miners also sold their wares directly to the users at other places, and in 1904 F J Snell noted that although most of the whetstones were sold to an agent in Honiton 'at the many of the fairs held in the neighbourhood, the whetstoners were their own merchants. At Tiverton fair they might be seen standing behind their wares, which were erected in a pyramidal form, and chaffering with farmers and others who came to buy'. A recollection of Percy Lane's was of regularly taking a cartload of whetstones to a Taunton ironmonger in company with the last known whetstone miner, John Rookley, presumably in the 1920s. The Revd E S Chalk, writing in 1910 when the industry was near its end, remarked that the best market for the whetstones of late had been in Australia.

THE MINING VILLAGES AND MINERS' DWELLINGS OF BLACKBOROUGH AND PONCHYDOWN

I am grateful to Richard Tamplin for much of the information in the following account which is drawn from his 2002 study of the settlements of Blackborough and Ponchydown and their buildings.

Blackborough and Ponchydown, once separate settlements, now

together form the village of Blackborough. It is unlikely that either would now exist, or if they did would not be so large, if it were not for the whetstone industry. Richard Tamplin writes that: 'They are, in terms of location, settlements built to serve the pits now in the woods above so that the ribbon of sporadic housing they form is a clear reflection of their functional and physical relationship to these former workings. Their layout, of large plots each containing one or a pair of cottages, can still be deciphered on the ground and on plans......The character of a late smallholder-miner settlement thus just about survives, most significantly in Bodmiscombe Lane where the original form, layout, plots and buildings create an almost untouched historic landscape of whetstone settlement.'

Richard Tamplin compared the Census Enumerators' Returns for 1841 with the Tithe Maps and Apportionments for Blackborough and Kentisbeare (compiled just a few years later, in 1844 and 1842 respectively) to find out which dwellings were occupied by whetstone workers. The Census Returns give the names of every person in a household and the occupation of the head of the household and of some other persons, but addresses are not often given. The Tithe Maps give the names of the occupiers and owners of every parcel of land (however, only leaseholders were treated as occupiers so that persons without a formal lease do not appear even though they may be living in the property). By comparing the two sources of information it is possible to find out which dwellings were occupied by whetstone workers, although the comparison is not exact because of the differences in dates of the sources. Richard Tamplin found that of the 36 dwellings within the area studied, at least 24 were occupied by whetstone workers, showing that in the early 1840s both Blackborough and Ponchydown were very much dominated by mine workers. The sizes of the gardens (up to ¾ acre) attached to whetstoners' properties suggest that many of the miners were also smallholders. Some of the cottages came with rights to work whetstones. For example, a 1915 sales particular of the Wyndham Estate (which owned much of Blackborough) notes that Post Office Cottage in Blackborough was 'let to Mr C Radford at 1s 4d per week and 5s per annum for the Scythe Stone Pit' (Devon Record Office 3223A/P28). The same family occupied this house at least as far back as 1841.

The question of whether Blackborough and Ponchydown originated solely as settlements to house whetstone miners is less certain. There appears to be evidence from the Tithe Apportionment for Blackborough (1844) of recent encroachments by dwellings onto land along the northern and western edges of the common, and several of these were occupied by miners, showing that Blackborough probably developed as a whetstone mining village largely in the early 19th century by the building of new dwellings to house whetstone miners. As far as Ponchydown is concerned, there is no evidence from the 1842 Tithe Apportionment of new intakes from adjacent land, but Richard Tamplin notes that four of the ten 'whetstone dwellings' were located just below the area of whetstone working and were probably intakes from the common at some stage. The whole of Ponchydown appears to be made up of cottages with large gardens enclosed from the large fields to the west. He thought that this might indicate that Ponchydown originated earlier than Blackborough. Certainly Ponchydown is marked on some 18th century maps, such as Down's map of 1765. Another centre was the Baptist congregation of Sainthill founded in 1803. It was a small community of about a dozen dwellings, less than a mile from Ponchydown. Robin Stanes noted that in the 1871 census eight whetstone miners were recorded in different families at Sainthill.

THE WHETSTONE MINERS

The miners had a reputation as a wild and rough people, but much of this was probably not based on any solid facts, and doubtless some writers have exaggerated their behaviour for dramatic effect. They were reputed to speak a strange dialect of their own, and were even thought to be Welsh or Cornish by some. R D Blackmore, the celebrated author of *Lorna Doone*, wrote a novel called *Perlycross* (identified with Culmstock) in which a Dr Fox treats an injured miner. The doctor recounts his meeting to a friend by saying 'You know they are a colony of Cornishmen', 'Yes and a strange outlandish lot, having nothing to do with the people around...'. Robin Stanes explains that their reputation was probably a natural consequence of their being a group of people living in a fairly isolated spot carrying out an unusual and dangerous living – the sort of reputation in fact often attached to small isolated mining communities. One of the centres of the community was the Ponchydown Inn which closed in 1963. They were said by the Revd E S Chalk not to welcome strangers in the inn.

The true character of the miners was occasionally maligned in fiction. For example, in the story in an American magazine of 1854

(*Harper's New Monthly Magazine*) which I have already mentioned, the portrayal of the miners is unflattering and they are described by the author as sickly drunkards, but much of this portrayal was for the dramatic effect of the story. The miners are described more flatteringly following an 1831 visit by Shirley Woolmer who wrote that 'The men behaved well, rationally replied to my interrogatories, and assisted me in procuring fossils'. She collected '....several clumps and groups of univalves and bivalves, small white nodules of different sizes, round as marbles; trigonia aliformis, fig-formed alcyonite, poppi-formed alcyonite, and lemon-shaped alcyonite.... the sand-stone containing the fossils was so damp, that with little exertion I could break it asunder to sort out the shells...' Visitors to the whetstone mines seeking fossils were probably quite common by the time of Shirley Woolmer's visit.

Robin Stanes noted that the census recorded 45 whetstone workers in 1841, 27 in 1851, 58 in 1861, 15 in 1871 and 6 in 1881. Because women were not included, the 1841 figure is likely to be nearer 70 and the industry seems to have been at its peak then, as the Revd Chalk recorded 24 pits in operation. There was a sharp reduction in the numbers of men working in the mines after 1871, probably owing to the concretions having been almost worked out. None of the figures for the numbers of workers are large considering that the industry served a countrywide and international market.

In Victorian times, when we have evidence from census returns, and probably for many years before, the industry was dominated by a small number of families. Robin Stanes notes that of 151 whetstone workers names recorded in the censuses, 31 were Radfords and 30 Bakers. The remainder, with only a very few exceptions had one of the following eight surnames: Bond, Pring, Coombe, Potter, Moon, Thomas, Rookley and Cording. This was clearly a small closed community amongst which intermarriage was probably not uncommon. The Radfords were particularly prominent in the community. In 1871 Charles Radford is described as 'Scythestone Cutter and Post Office Master' and continued in this role until at least 1910. The same property in Blackborough (formerly Post Office Cottage, now called 'West View') was occupied in 1841 and 1851 by Isaac Radford, described in the census for those years as 'Whetstone labourer' and 'Whetstone cutter' respectively. The miner shown in photo 4.16, on page 72, is believed to be a Mr Radford.

The organisation of the miners into gangs of four men has been described above (page 74). The work was very much a family affair, closed to 'outsiders'. Workers were recorded in the censuses as scythestone 'cutters', 'polishers' and 'workers'. There was also a category of worker called 'sand drivers'; the latter job, which was the removal by wheelbarrow of waste rock and sand, was often carried out by boys and girls.

A memory of one of the last whetstone miners is recalled by the Revd R S Chalk. He writes that 'I remember my Father [the Revd E S Chalk] taking Hal Gundry [Squire of the Grange, Broadhembury], with my brother and myself, to visit Jas. Rookley's pit about 1917-18. No one was working at the time, so we did not go inside. The site was on North Hill'. He continues: 'On a later occasion (perhaps about 1925-26) a friend and I were walking over North Hill when we heard a rhythmic sound, as of a pick-axe, below ground not far from us. We realised it was the last scythe-stone worker still plying his trade. Naturally we were very careful not to walk over his head or anywhere nearer, knowing how soft was the green-sand soil'.

Accidents and the health of the miners

Mining is a dangerous business at the best of times, but the whetstone miners were tunnelling into strata that were treacherous and needed continuous support. The unconsolidated or weakly cemented sands forming the roof and sides of the tunnels were particularly prone to sudden catastrophic collapse. We have no figures for the numbers of accidents and deaths in the whetstone workings, but the Revd Chalk says that a life a year was lost. Apart from the dangers of the actual workings, the whetstone workers' health seems to have been affected by respiratory diseases caused by the inhalation of dust both underground and while shaping the whetstones. In 1831 Shirley Woolmer visited the whetstone mines and commented on the complexion of the miners: 'The fine ruby complexion of the youths employed in excavating the earth excited my surprise, as it exceeded the usual flush of nature; also as I stood at the mouth of the cavern, I saw a tall, slender old man, coming out of the gloomy recesses, whose visage was a light carmine, the colour probably the effect of some peculiar essence arising from the bowels of the earth'.

William Fitton, in his 1836 account, quoted from a letter of 1813 relating to the health of the miners: '...the preparation of the sithe-stones is so injurious to the health of the work-men, that in the little village of Punchey Down, which is inhabited exclusively by the stone-cutters, the writer saw but one elderly person, and was informed that few reached the age of forty. The complexion and figure of the greater number he describes as striking and interesting in a high degree; but it was the hectic aspect, and itself a proof of disease'. Curiously William Fitton wrote that when he visited the mines in 1825 he did not recall anything that suggested ill-health among the miners. The Revd E S Chalk says that many men did not reach the age of 50, having died from inhalation of the fine dust or 'smeech' produced during the process of dressing the stones.

The Revd Kirwan, in his 1866 account, notes of the work of the whetstoners: 'It is also extremely injurious to the workmen. I believe that none of those who are engaged in it ever reach the term of life allotted to man – three score years and ten: few indeed exceed the age of forty. The men appear strong and healthy, and have a stalwart appearance, but in walking with them over the hill I have observed that they are very soon out of breath and quickly become fatigued. This ruddy aspect is but the hectic flush – itself a sign of a disease. The injurious nature of their occupation arises from the fact that minute particles of sand being set free, these become diffused in the air, and are thus drawn into the lungs, whereby asthma and pulmonary diseases are superinduced.' He goes on to note that the women also suffer health problems from their work in shaping the whetstones. 'Chronic rheumatism is the frequent penalty of this unfeminine labour, and if the women ply their task in the same room or shed as that in which the men chop the whetstones, they too are the victims of the same pulmonary maladies as the men'.

Robin Stanes has looked at the health of the miners in more detail in his account of the industry. He examined local death certificates and confirmed that lung disease was indeed the main cause of death among the workers: of 15 examined 13 died of some type of lung condition. It seems from Robin Stanes' analysis that many of the men died of a combination of tuberculosis and silicosis. Silicosis is caused by the inhalation of fine sand particles penetrating the lungs and can cause death by respiratory failure. It also increases the risk and severity of tuberculosis. The red faces of the miners referred to in Victorian accounts were either caused by the early stages of active tuberculosis or due to active silicosis, or more probably a combination of the two. The women who rubbed and polished the stones must have also been exposed to dust, but the fact that the death rate was lower among the women may be due to the fact that the process was carried out using water and in areas with greater ventilation than within the mine tunnels.

The last whetstone miner is believed to have been John Rookley, who is perhaps pictured at the mouth of a tunnel in photo 4.15. It is curious to think that there must have been a particular day, probably sometime in 1929, when he walked out of his whetstone tunnel for the last time, bringing to an end an industry that had a history of over 200 years. In the evocative words of Derrick Rugg:

'.....the industry faded away until the last miner shrugged on his jacker, and wistfully, one imagines, laid aside his hammer so that the peculiar music of Blackborough was carried away on the wind and mist, growing fainter, till all force was spent – save the whispering held by memory'.

Chapter 5
'CATHEDRALS OF STONE':
The Beer Stone Mines of East Devon

INTRODUCTION

We now move in our journey of exploration from Blackborough to southeast Devon where, about a mile west of the village of Beer, we find the fascinating Beer Stone mines which the celebrated landscape historian W G Hoskins called 'one of the most exciting things in Devon' (for their location see the map, **1.1**). The area lies within the East Devon AONB (Area of Outstanding Natural Beauty). A special and unusual development of the Chalk, of Cretaceous age, formed about 90 million years ago, has been mined for building stone from underground tunnels for about 2,000 years. Today the old underground workings, called Beer Quarry Caves, are open as a tourist attraction, and there are regular underground tours during the season.

Perhaps the best way to appreciate the geological setting of the Beer Stone is to take a boat trip from Beer or from Branscombe Beach. Details of trips available can be found on the Jurassic Coast website www.jurassiccoast.com. From a boat there are spectacular views of the coastline, including the soaring cliffs of Beer Head, the most westerly Chalk cliffs in England. From Beer Head westwards towards Branscombe Mouth you will see not only the continuation of the high Chalk cliffs, but the Upper Greensand beneath which emerges owing to the easterly tilt of the strata. The division between the Upper Greensand in the lower part of the cliff and the Chalk in the upper part can be seen in the photo of the cliffs (**5.1**). Alternatively, walk from Branscombe Mouth eastwards along the base of the cliffs towards Beer Head (keeping well clear of the cliffs owing to the risk of falling rocks).

Another spectacular feature of the coast here is the presence of a great landslip – the Hooken Landslip. The last major slip took place in March 1790 when a great crack opened up overnight in the cliff, and huge blocks of Chalk and Upper Greensand over 1/3 mile long

5.1. The Upper Greensand and Chalk cliffs west of Beer Head. The white Chalk in the upper part of the cliffs can easily be distinguished from the underlying yellowish-brown Upper Greensand.

5.3. The Beer Stone adit in Hooken Cliff.

5.2. The Hooken Landslip. Huge masses of Chalk and Upper Greensand broke away and subsided towards the sea in March 1790. The mouth of an adit at the Beer Stone level can just be made out (see photo 5.3 for a closer view).

slid down towards the sea by up to 260ft. The result is a jumbled mass of broken Upper Greensand and Chalk, now very overgrown, called Under Hooken (photo **5.2**). In the back face of the landslip can be seen an adit mouth which marks the position of the Beer Stone in the cliff face (photo **5.3**) and shows that at some time an attempt has been made to work the Stone here (more details are given on pages **95 and 97**).

The stone was first worked by the Romans and we can speculate about how it might have been first discovered. About 2,000 years ago, perhaps a galley commanded by a Roman crew was sailing along the coast. One of the officers looked up at the great white cliffs, which may have reminded him of the monumental white stone used in buildings in Rome. A party was sent out to scramble up the cliffs to examine and report on the suitability of the rocks for use as building stone. The Upper Greensand would have been unsuitable,

as was most of the chalk, but the searchers reported that halfway up the cliff was a layer about 13ft thick of homogenous creamy white stone. On return to the mainland a search inland would have found the same stone on the side of a valley, in a more easily worked position. A decision was taken to burrow into the hillside – the stone was found to be of good quality and eminently suitable for ornamental stone work – and the history of the Beer Stone mines began.

DEVON IN THE LATE CRETACEOUS 90 MILLION YEARS AGO

What was the world like when the Beer Stone was formed? Following the period of whetstone formation in the Upper Greensand about 100 million years ago, described in the previous chapter, sea levels continued to rise and eventually an extensive shallow sea lay across much of Europe. The Dartmoor Island, described in the last chapter as standing up from the Upper Greensand sea, was soon to be drowned by the rising waters, and in fact only parts of Wales and Scotland remained as land in the Chalk seas. On the floor of the ocean, which was probably between 330 and

1640ft deep, accumulated the unique and curious deposit that we now know as chalk. It is almost entirely made up of the remains of tiny marine plants (algae) that lived and floated in the surface waters of the ocean. The algae, called coccolithophores, formed a kind of ball around themselves, made up distinctive tiny circular plates called coccoliths. When the algae died, the plates separated and fell in their countless billions, like snow, to the sea floor. Also, some of the algae were eaten by copepod crustaceans which then deposited the coccoliths on the sea floor as faecal pellets. If this theory is true, part of the chalk is made up of the droppings of these animals – a curious and somewhat repellent thought! The coarser-grained fraction of the chalk (0.01 to 0.1 millimetres) includes fragments of other fossils, including foraminifers, ostracods, echinoid plates and fragments of bivalves. Clay and silt form only a very small proportion of most chalk.

Apart from the microscopic coccoliths, the Chalk seas teemed with life. Sponges were abundant on the sea floor. Sea urchins and worms burrowed into the sea bed. Bivalves and brachiopods lived on the sediment of the sea floor. In the waters above, fish swam. Ammonites, so abundant in the older Jurassic seas, floated in the Cretaceous oceans, but were not so common.

The geology of the Chalk and the Beer Stone

We often think of chalk as very much a homogenous rock showing little variability and laid down steadily on the sea floor. However, when it is looked at more closely it can be seen to be quite variable in detail, with varieties such as: chalk made up of shell fragments larger than sand grade; gritty chalk made up of sand-sized fossil fragments; nodular chalk; and chalk with fine laminations. Chalk was not laid down steadily on the sea floor and there were many gaps in its formation, sometimes lasting hundreds of thousands of years. The gaps are marked by so-called 'hardgrounds'. Nodular chalk was formed by the selective hardening of soft chalk just beneath the sea bed.

In the Beer area, the Chalk is superbly exposed in the sea cliffs around Beer Head and Beer village (see photo 5.1). By studying and measuring the sections in detail over the years, geologists have produced a representative section through the Chalk. The old subdivisions of the Chalk - Lower, Middle and Upper - are no longer used, except informally. Interested geologists have discussed ways of dividing up the Chalk, and a new set of names has been proposed.

The main new subdivision is into a lower 'Grey Chalk Subgroup' (approximately equivalent to the old Lower Chalk) and a 'White Chalk Subgroup' (approximately equivalent to the old Middle and Upper Chalk combined).

Immediately over the Upper Greensand at Beer lies a distinctive bed, varying between 1 and 41ft thick, which was formerly called the Cenomanian Limestone, but is now known to geologists as the Beer Head Limestone. Subdivisions lettered A, B and C have been recognised within it (see 5.4). Remarkably, this generally rather thin bed is the condensed equivalent of the whole of the Grey Chalk ('Lower Chalk') of southeast England which is there on average 300ft thick.

The 'Middle Chalk' of the Beer area is roughly equivalent to the new subdivisions of Holywell Nodular Chalk Formation and overlying New Pit Chalk Formation. As the name suggests, the Holywell Nodular Chalk is made up largely of hard nodular chalks, but its main point of interest for us is that it contains within it the distinctive bed of limestone called the Beer Stone that we shall look at in more detail shortly. The New Pit Chalk is blocky white chalk with marl seams. The 'Upper Chalk' at Beer consists of Lewes Nodular Chalk - hard nodular chalks and hardgrounds with flint

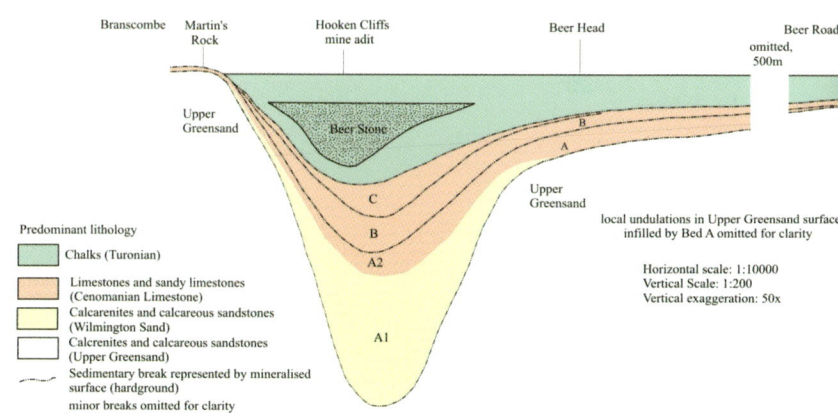

5.4. *This diagram through the Hooken Cliff shows how the Beer Stone relates to the rest of the Chalk of the Beer area. Note how the Beer Stone has a restricted extent and dies out rapidly to east and west.* Courtesy of Dr R W Gallois.

seams – and the overlying Seaford Chalk which is firm white chalk with conspicuous tabular and nodular flint seams. The sequence of strata in the Hooken Cliff and adjacent areas is given in the diagram (5.4), which shows particularly the position of the Beer Stone.

Flint at Beer

The Beer area is also well known, particularly by archaeologists, as the source of high-quality flint that occurs as layers in the Chalk, and which was used in prehistoric times to make stone tools. Although, as I mentioned in Chapter 1, flint has been mined in parts of England (for example at the celebrated Grimes Graves in Norfolk), it is not known for certain that there were any mines in the Beer area. Instead the prized flint was probably mostly dug out from cliff exposures, although the possibility of shallow excavations cannot be ruled out. There has been a general tendency for some archaeologists to assume that the flint used by prehistoric people in Devon was dominantly sourced from Beer, but John Newberry has pointed out that geologists have recorded numerous inland sources of flint in Devon, many of which could have provided flint of high enough quality to make excellent tools.

THE BEER STONE

As we have seen above, the Beer Stone occurs near the base of the Holywell Nodular Chalk. On the coast it forms a very locally developed lenticular unit, thickest in the Hooken Cliff area, but thinning out rapidly to the west and east, as can be seen on the diagram, 5.4. It is not present at Beer Head to the east of Hooken Cliff, or in Branscombe East Cliff to the west of Hooken.

Unlike the nodular chalks, the Beer Stone is a distinctive creamy-white bed of limestone. Although it has a few patches which have been disturbed by burrowing animals, and contains a few fossils, it is mostly homogeneous in texture and grain size, and is the only true freestone in the Chalk of Britain (a freestone is a type of building stone which can easily be cut and worked in any direction without breaking). It is up to maximum of about 13ft thick and can be broadly divided into two parts. The upper, about 7 or 8ft thick, is thick-bedded and the lower, about 5ft thick, is harder and thinner-bedded. Just above the top of the Beer Stone is a layer which the miners called the 'roof course' or 'cockly bed', consisting of about 3ft of nodular chalk.

Henry De la Beche, writing in 1826, gave an interesting description of the detailed subdivisions of the Beer Stone recognised by the quarrymen: 'The workmen enumerate the following beds, beneath 18ft of indurated chalk without flints, which they call *Skull*.

1. Upper bed, - eighteen inches thick, rather hard.
2. Eighteen inches thick, soft.
3. Two feet thick, rather soft.
4. Two feet thick, rather soft.
5. Sixteen inches thick, hard.
6. Sixteen inches thick, mixture of hard and soft.
7. Eight inches thick, very hard.
8. Two feet thick, very hard.

[Total thickness 12 feet 4 inches]

Beneath these are about 5 or 6 feet of a hard white calcareous rock, which burns into good lime, but is not quarried as the others are. Beneath this the sandstones [Upper Greensand] commence'.

The upper thick-bedded division referred to above is equivalent to the beds numbered 1 to 4 by De la Beche, and the lower harder division is equivalent to his beds 5 to 8. Sections through the Beer Stone drawn on a map of 1873 (illustration 5.7, page 89) show a subdivision into five units, which probably represent natural variations (beds) within the rock. If these beds are lettered A to E, in order from bottom to top, their possible equivalence to the beds numbered by De la Beche in 1826 is as follows: A = bed 8, B = beds 7 and 6, C = bed 5, D = beds 4 and 3, and E = beds 2 and 1.

Earlier descriptions of the Beer Stone, for example by Alfred Jukes-Browne in 1903, usually characterise it as a crystalline granular limestone composed mainly of small fragments of the broken-up shells of the bivalve *Inoceramus*. Jukes-Browne wrote that 'The Beer Freestone consists almost entirely of coarse calcareous shell-fragments. Some of these are prisms of *Inoceramus*, and others are plates of *Echinoderms*, but the larger numbers are fragments of irregular shape, which show a peculiar perforated or cellular-like structure. There are no Spheres and very few Foraminifera, but the interstices between the shell-fragments are filled with amorphous calcareous material, the whole being in a semi-crystalline condition. The rock is therefore a shell limestone, and can hardly be called a "chalk"'.

We are still not sure about the exact composition of the Beer

Stone, and some geologists think that it is made up mostly of broken fragments of micro-crinoids (sea-lilies), a type of animal allied to sea-urchins. The Beer Stone may have formed on a somewhat shallow sea-floor, perhaps bounded by north-south-trending faults on either side. On the sea floor, swept by currents, may have grown a forest of colonies of micro-crinoids. Breakdown of these may have produced a deposit composed mainly of sand-sized grains made up of parts of the broken crinoid columns. The theory that the Beer Stone is made up largely of fragments of the bivalve genus *Inoceramus* seems less likely in view of the durable nature of the large *Inoceramus* shells. It would take a very high energy environment – perhaps affected by storms – to be able to break the thick and sturdy *Inoceramus* shells into sand-sized particles.

Interestingly, when the Beer Stone is dissolved in acid and the residues studied it is found that they contain many minerals which can only have been derived from the Dartmoor Granite. This suggests that Dartmoor might have still stood up as an island at this time, just at the beginning of a period when the Chalk seas began to deepen.

THE HISTORY OF BEER STONE MINING

We know that the mines have been worked from at least Roman times, for Beer Stone has been found in a 2nd to 4th century villa at Seaton, and in buildings in Exeter. The Romans were established in southwest England at two main centres - Dorchester and Exeter. About midway between these towns is Seaton, where the harbour at Axmouth was an important Roman port, and it is possible that Beer Stone was exported from there. The Saxons also used the stone, which can be seen in the beautiful Saxon cross in Colyton Church.

Following the conquest in 1066 the Normans began a major building programme of castles, churches, cathedrals and manor houses, stimulating a large demand for stone from the mines. Beer Stone was used in at least 24 of the 44 great cathedrals of the Norman period. Exeter Cathedral is a well known example of its use (see page 104), and the stone became widely used during the 13th and 14th centuries. It was employed in such diverse and important buildings as the Tower of London, Westminster Hall, London Bridge and Winchester Cathedral.

Much information about the use of Beer Stone in Exeter Cathedral is contained in the Cathedral Fabric Rolls. Excerpts from these are shown on a display board in the small museum at the entrance to the Beer Quarry Caves. The accounts of the fabric of the cathedral for the period 1279-1353 have been translated and edited by Audrey Erskine, and there are numerous references in the accounts to the Beer Stone quarry. The earliest recorded purchase date for Beer Stone to be used in Exeter Cathedral is 1302-1303, although it is possible that the stone may have been used for cathedral purposes for many years before this date. Philippe Planel notes that many of the cathedral accounts refer to the costs of transporting the stone, and he quotes a typical entry in 1318: Christmas term Week 5, 1318 'For carrying 5 great stones from Beer Quarry to Exeter'. Many other entries refer to larger consignments by land and sea, some involving over 100 tons of stone. Other separate accounts are for carriage of stone by sea from Beer quarry, and in the period 1279-1353, these totalled thousands of tons. A characteristic entry for Easter term Week 11, 1325, records 'Item 3 barges of stone from Beer and Salcombe quarries 24s 6d'. John Torrance writes that a quarry contractor called William de Galmeton or Galmyngton provided stone to the cathedral in the 1330s and 1340s. In 1341 accounts show that as well as stone, he delivered two loads of lime to the cathedral, indicating that lime as well as freestone was then being produced at Beer. W G Hoskins refers to 32 cartloads of stone bought by Exeter Cathedral in 1429-30.

Following the virtual completion of Exeter Cathedral in the 15th century, there was a period of renewed church building and improvement of parish churches in Devon and elsewhere, and the mines reached their peak in the 15th and early 16th centuries. Beer Stone was used mainly for decorative work inside churches and is found in most east Devon churches. Such was the cost of transport that churches farther away, say more than 30 miles from the quarry, were only able to use Beer Stone if the church had a rich benefactor able to support the cost of transport.

However, with the sudden end of church building in about 1540, following the dissolution of the monasteries by Henry VIII, there was an inevitable decline in the output of the mines, particularly since Beer Stone is not generally suitable for outside work. Consequently the mines were unable to take part fully in the period of country house building which succeeded that of ecclesiastical building. For about 300 years from the mid-1500s, John Scott notes that the mines were worked only spasmodically, mainly to supply stone for the building of houses. Notable amongst those buildings in which Beer Stone was used during this period are St Paul's

Cathedral, the portico of the Guildhall in Exeter (which is of Elizabethan age), and Shute House near Colyton (originally built in 1550 and remodelled in 1787-90).

The Walrond family, whose seat was at Bovey House, were owners of the Beer Quarry Caves and lords of the manor from 1300, and in 1550 they rebuilt Bovey House entirely from Beer Stone. In 1772 the mines passed by marriage (of John Rolle to Judith Walrond) to the Rolle family who have owned them ever since.

Philippe Planel has examined leases in Rolle estate documents held at the Devon Record Office and reported on them for the Devon Historic Environment Service. He noted that quarrying of the Beer Stone and lime-burning often went hand-in-hand, as illustrated by the Rolle leases and the Ford family papers. The Fords were a notable local family of lime burners who were active in the Branscombe area and then later extended their activities to Beer in the early 19th century when they also began to quarry the Beer Stone at the New Quarry (see page 94 below). Philippe Planel noted that the exact location of pits, quarries and lime kilns mentioned in documents are often difficult to determine. The tithe map of 1840 shows many plots with 'quarry' names which relate to both Beer Stone and to extraction of limestone for burning.

Two finely written 17th century leases of 1664 and 1674 are in the Devon Record Office (96M/Box 35/3). The earlier lease of 1664 is between Edmond Walrond of Bovey House and Robert Starr of London, a merchant. It refers to 'all his quarrie and quarries of Freestone....commonly called Bear Quarrie, and Beer Quarrie hills'. The annual rent was £5. The agreement specified that the lessee should '...leave sufficient pillars att all convenient places for the support and maintenance of said quarry', thereby proving that at least some of the workings were underground.

Richard Polwhele, writing in 1797, included a brief description of the Beer Stone. He noted that: 'When hewn out of the quarry, the freestone of Beer cuts as soft as the Bath Stone, which it greatly resembles. It does not usually stand the weather; especially in exposed situations. Yet I have seen some pieces of it, which have been a long time in buildings without crumbling. It is very good for inside work: all the vaulted roof and ornaments of the arches at the cathedral of Exeter are made of this stone. It bears the heat of the fire better than any other sort of stone, and is therefore useful for hearth-stones'.

Charles Vancouver, in his 1808 survey of the agriculture of Devon, mentioned the Beer Stone which he said is 'in every respect equally valuable' as Portland Stone. The Lysons in 1822 mentioned the Beer quarries only in passing and noted that 'A considerable quantity of it is sent coastwise'. Henry De la Beche, writing in 1826, noted that 'The quarry is interesting as a cavern, and affords a miniature representation of the Maestricht quarries; the roof, which is nearly even and parallel to the floor, being supported by large square pillars. The depth of this cavernous quarry from the entrance is about 170 yards'.

As we have seen, working of the mines was intermittent until the great Victorian revival in church building and restoration gave another boost to production. The old workings (the Old Quarry or Beer Quarry Caves) on the south side of Quarry Lane were replaced from the late 18th century by a quarry and mine on the north side (the New Quarry) and very little stone was taken from the old workings; the last known removal of small quantities of stone was in the 1920s. Beer Stone was extracted on a small scale on demand from the New Quarry until about 2003.

THE BEER STONE MINES

The Beer Stone mines are located about a mile west of the village of Beer. The oldest workings [SY 215 894] are situated south of the lane (Quarry Lane) and are called the Old Quarry or Beer Quarry Caves. The more recent New Quarry workings [SY 215 898] are on the north side of the lane. The surface workings and associated buildings are shown on the 1:2500-scale Ordnance Survey map of 1887 (5.5). The presence of an adit and the record of a shaft suggest that unrecorded Beer Stone workings may well lie under Hooken (see page 97 below).

Beer Quarry Caves

Two plans of the underground workings at Beer Quarry Caves are reproduced here (5.6 and 5.8). The first survey dates from 1873, and the second is a plan made in 2008. The labyrinthine nature of the tunnels is clear from a glance at the surveys. The underground workings surveyed in 2008 cover an area of about 2½ acres but there may be unsurveyed workings of uncertain extent. Some sources refer to the existence of eighteen closed entrances to the caves, but more work is needed to establish their location and number.

A large (about 3ft by 4ft) interesting plan on a scale of 30ft to 1 inch, dated October 1873, is held at the Clinton Devon Estates

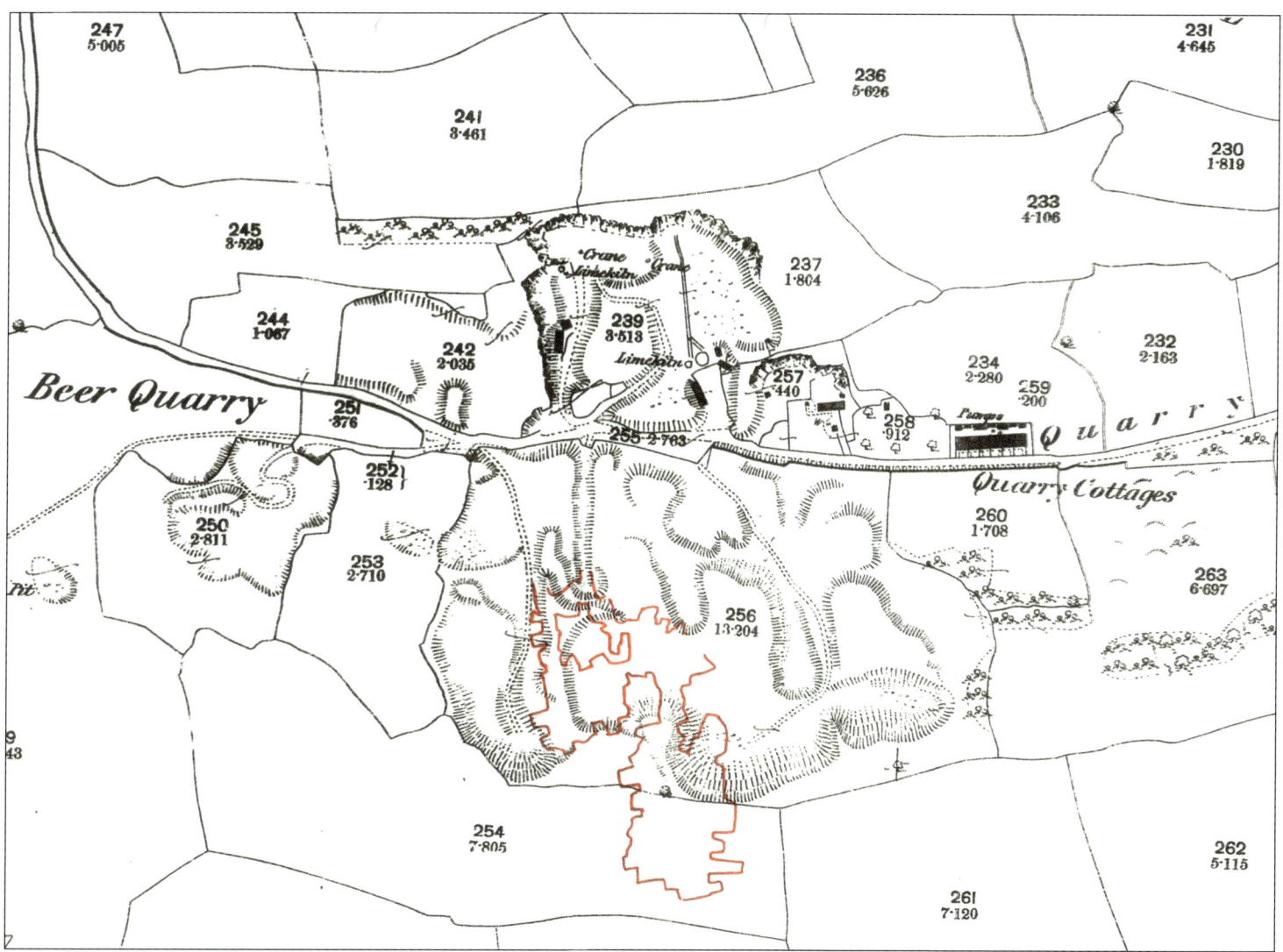

5.5. *Part of an Ordnance Survey 1:2500-scale map of 1887, showing the workings north and south of Quarry Lane, Beer. The modern survey by Chris Wood (see 5.8) has been superimposed on the map in red to show the position of the underground workings of Beer Quarry Caves in relation to surface features. Not to original scale.*

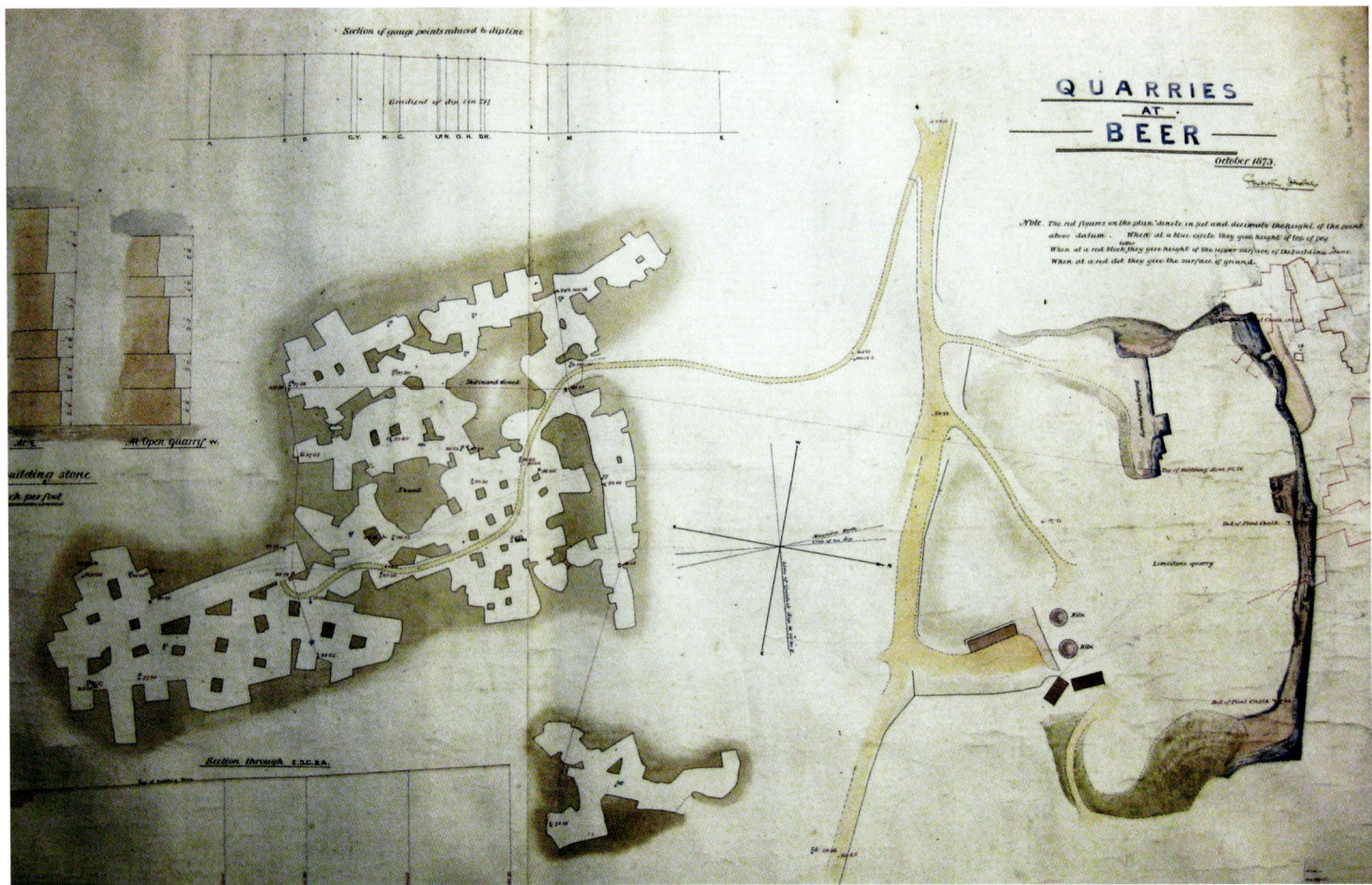

5.6. *Part of an 1873 plan of the Beer Quarry Caves and the New Quarry north of the lane.* Reproduced by courtesy of Clinton Devon Estates.

Office, and shows the quarries to the north (New Quarry) and south (Beer Quarry Caves) of Quarry Lane. It was surveyed by Sheriton Holmes, and part of it is reproduced in **5.6**. Sheriton Holmes was a mining engineer from Newcastle on Tyne, and John Scott has letters written by him showing that he travelled to Seaton by train, carried out the survey and returned to Newcastle, and within a week the completed plan had been sent to Beer! His name can be found inscribed on the wall of one of the underground tunnels.

The 1873 plan shows details of the underground workings south of Quarry Lane in the area of what is now called Beer Quarry Caves. The method of working was by pillar and stall, the roof being supported by pillars of unworked stone. The plan shows a major area of connected workings, but there is another small area of workings to the east, not obviously connected to the main workings. Chris Wood notes, however, that there is a possible link between the main workings and the smaller system to the east via a low passage on the eastern side of the exit tunnel. He also notes that there does not seem to be any indication that the workings extend farther south beyond the tour caves, the southern end of the workings being solid stone. Chris Wood also points out that there is a substantial amount of backfill thrown in from the quarries in the hill above, potentially suggesting that there may be workings which are now buried. However, if this was the case, the backfill would have had to be mostly in place when the 1873 map (**5.6**) was drawn.

The large 1873 plan also shows several sections through the Beer Stone which is between 11ft 1in and 13ft 1in thick (**5.7**). Interestingly, the part of the plan showing sections through the Beer Stone suggests that the stone was worked in five individual units, which probably represent natural divisions (beds) within the rock. On page 84 above I have suggested possible correlations between these five units and the eight beds recorded by De la Beche in 1826.

Part of the underground workings of the Beer Quarry Caves was surveyed in 2008 using 3-D laser mapping techniques. The survey was carried out by Exeter University (Camborne School of Mines) for

5.7. Sections through the Beer Stone showing five subdivisions (see page 84). This is a detail from a large plan of 1873 held in the Clinton Devon Estates archive (see 5.6). Reproduced by courtesy of Clinton Devon Estates.

the East Devon AONB (Area of Outstanding Natural Beauty) Partnership working with the Historic Environment Team of Devon County Council. The actual survey work was done by Chris Wood, then at Camborne School of Mines, who has kindly produced a simplified version of the original survey for use in this book. The resulting plan (**5.8**) is broadly similar to the 1873 plan, but closer inspection shows that there are many differences in detail. The southernmost workings seem to have been expanded considerably to the west since 1873. The northern workings also differ in detail from the 1873 plan. The small area of workings to the east shown on the 1873 plan was not surveyed in 2008. In **5.5** the modern plan of the underground workings is shown superimposed onto the 1887 1:2500-scale Ordnance Survey map to show the position of the workings in relation to surface features.

The Beer Quarry Caves are a popular tourist attraction open for visitors seven days a week from the Monday before Easter to 30th September from 10am-5pm (last tour) and in October from 11am to 4pm (last tour). These opening times were correct in 2009, but can be checked by phoning 012977-20986, or see www.beerquarrycaves.fsnet.co.uk. There are conducted tours of the underground workings lasting about an hour. During the winter the Caves are closed because they are the home of eight protected species of bats (see page **105**, below). The temperature in the caves is a constant 12° C (55° F) all year round. Your first port of call as a visitor is the visitor centre where refreshments are available and there is an exhibit and archive photographs of the workings. An interesting booklet entitled *Out of the Darkness* by John Scott and Gladys Gray describing the workings is available, and I have drawn, with permission, on its contents for parts of this account.

The caves were very nearly lost as a heritage attraction. In 1983 a man became lost for 16 hours in the maze of workings, and this incident led to a proposal by the lessees of the quarry to close the quarry entrance to prevent a recurrence of this event. They were saved by John Scott, a local man who had known the caves since childhood. He sought permission from the freeholder, Lord Clinton, who agreed to give John a lease to keep the caves open; they have been a well-known tourist attraction since then.

After passing through the visitor centre, where you will be issued with a hard hat, you approach the entrance to the caves by a short woodland path. There are two openings (photo **5.9**): the one directly ahead is the exit at the end of the guided tour and that to the right

*5.8. A modern plan of part of the Beer Quarry Caves workings, based on a survey by Chris Wood of Camborne School of Mines in July 2008. The survey was carried out by Exeter University (Camborne School of Mines) for the East Devon AONB Partnership working with the Historic Environment Team of Devon County Council. This version was prepared for this book by Chris Wood. Parts of the workings are not surveyed. Compare with the 1873 plan (**5.6**). The position of the underground workings in relation to surface features is shown on map **5.5**.* Reproduced by permission.

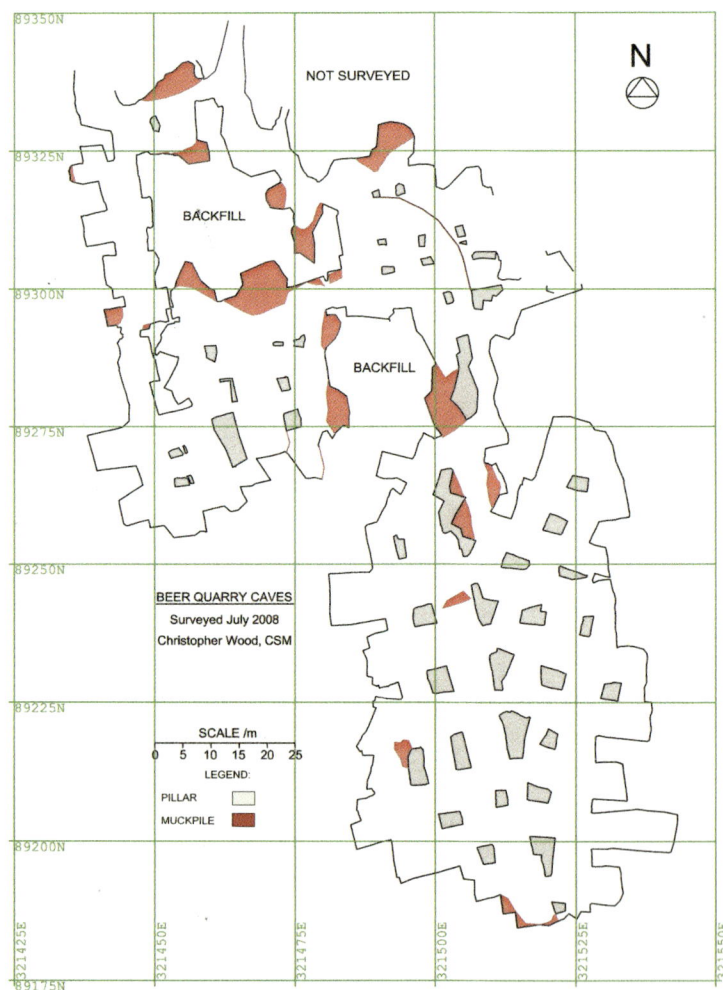

5.9. *The entrances to Beer Quarry Caves in 2009.*

progressively younger in age. There are workings believed to be of Saxon, Norman, and Medieval and later ages. The age of the tunnels can sometimes be judged by the different styles of working, but later alterations such as pillar robbing (page **98**) have often blurred the distinctions. The Roman workings tend to have rounded arches (photo **5.10**) while the Saxon and later workings are generally square with flat-topped chambers (photo **5.11**).

Roman workings

The entrance by which visitors enter the caves is of Roman age, and Roman coins and tools have been found around the entrance. The Roman workings have rounded arches (photo **5.10**) and the tool marks of the Roman quarrymen are visible on the walls of the workings. Despite the large quantities of stone removed in Roman times, it is a puzzle that there are only a few authenticated examples

5.10. The Roman entrance to Beer Quarry Caves, showing characteristic rounded arches. Photo © Beer Quarry Caves.

leads into the area of earliest (Roman) workings in which there is a small museum. Formerly there were several entrances to the Caves (some unsubstantiated sources mention as many as eighteen), but most have been filled in. Recently a third entrance to the Caves has been re-opened by Natural England to increase the area of workings available to bats. In the museum area there are a number of display boards with information about the industry. One of the boards shows extracts from the Exeter Cathedral Fabric Rolls. The quarry was owned by Exeter Cathedral in the Middle Ages, and a display board shows how Beer Stone was used in the construction of the cathedral. The museum has examples of the tools used to work the stone, and examples of worked stone including a large medieval window tracery rescued from Colyton Church and re-assembled here.

Visitors assemble in the museum area and the guide gives a brief introductory talk before the party proceeds into the workings. As the tour progresses we pass farther into the caves which extend to about 200 yards from the entrance. The Beer Stone forms a roughly horizontal layer generally between about 12 and 13ft thick, so that all the passages are on the same level. Since the Romans were first to work the stone, naturally the first workings that you enter are of Roman age, and then deeper in the caves the workings are of its use; the best known example is the 2nd to 4th century Roman villa at Honeyditches, near Seaton. There, a detached bath house was built of blocks of local chert with Beer Stone quoins. It may be that there are Roman buildings still to be discovered in the area, or that the stone was transported farther afield (to Exeter for example). It is also possible that stone was exported to other parts of the Roman Empire through the Roman port at Axmouth, near Seaton, but it would need considerable further research to establish this.

Saxon, Norman, medieval and later workings

After the Romans left Britain in the 5th century, the quarries continued to be worked, by the Saxons. The Saxon and later workings found deeper in the caves are distinguished from the Roman workings by their square outline (photo **5.11**). John Scott notes that the size of the workings indicates that the Saxons made limited use of the Beer Stone.

The great expansion of building during the Norman period meant that Beer Stone was in great demand for use in cathedrals, churches, castles and manor houses, and there are extensive workings of that age. John Scott notes that the workings of Norman and later age are apparently in side tunnels which lead off from a long straight passage of Saxon age. Because of the demand, the main working was

pushed deeper into the hill and galleries were driven to the east and west. The roof was supported at intervals by large unquarried stone pillars which give the striking appearance of galleries and chambers which the visitor sees today (photo **5.12**), and which is sometimes compared to the lofty feel of a subterranean cathedral. Working of stone continued in medieval and later times and there are probably extensive areas of the Beer Quarry Caves which date from these periods, although precise dating is difficult. Workings of possible medieval age are shown in photo **5.12**.

The full height of the workings can be seen only in a few places, since in most of the passages the floor level has been built up the dumping of quarry waste, or 'gob' as it was known to the workers. In the small museum at the entrance to the Caves can be seen a small handcart used for transporting quarry waste, known as a 'gobcart'.

In places the roof and walls of the chambers are crossed by cracks – these are either faults where some relative movement of the rock to either side has taken place, or joints where no relative movement has occurred. Some of the faults are lined with red clay which may be derived from material from the Clay-with-flints (see page **64**) percolating down from near the surface many feet above. Some of the areas of clay fill may be the lower parts of pipes of Clay-with-flints descending from the surface deep into the Chalk. In a visit on 26

5.11. Workings of possible Saxon age in the Beer Quarry Caves, showing characteristic square outlines. Photo © Beer Quarry Caves.

5.12. A characteristic view of the impressive interior of Beer Quarry Caves showing workings of possible medieval age. Photo © Beer Quarry Caves.

5.13. In the innermost part of the Caves, about 200 yards from the entrance, are the last areas of stone to be worked, where partly excavated blocks can be seen. Photo © Beer Quarry Caves.

September 1859 the Sidmouth antiquarian Peter Orlando Hutchinson noted in his diary that the miners tested the cracks or fissures in the mine by rubbing clay into them; if it did not crack with time, the fissure was not increasing in size. The miners used the red clay found in the fissures for holding candles in place. Candle ledges fashioned in the rock walls can also be seen. It is said that the quarrymen were allowed 5 candles a day and had to pay for them, but masons were allowed a free supply of candles. This no doubt was a source of dissent between the two groups of workers.

At the farthest extent of the workings, the face shows partly excavated blocks where the last stone to be removed was worked (see photo **5.13**).

For a time, beginning in the 1940s, the Beer Quarry Caves were used for the cultivation of mushrooms, and remains of some of the beds can still be seen. The Caves were opened to the public as a heritage attraction on 16 April 1984 and are visited by over 20,000 people a year.

New Quarry

New Quarry is situated on the opposite side of the lane to the entrance to the Beer Quarry Caves. The quarry was already being worked in 1774, and the plot is marked 'Quarryhill and barn' on an Estate Map of that date in the Clinton Devon Estate archive. It is shown on the large plan of 1873 (**5.6**, page **88**) where it is labelled 'Limestone [Chalk] Quarry'. At that date the quarry was probably used mostly for lime production and two circular limekilns are marked. On the west side of the quarry a small area is labelled 'Building stone quarry', but there are no obvious underground workings in the Beer Stone shown in this area of the 1873 plan. Underground passages (see photo **5.14**) from which Beer Stone has been worked are reported (according to the 2004 Devon County Minerals Local Plan) to extend from the north face of New Quarry for about 875 yards, beneath land between Bovey Coppice and The Rookery. The workings were by pillar and stall methods. I have not been able to locate any plans so it remains unclear whether such reportedly extensive workings actually exist. According to W G Hoskins, in his book on Devon, the New Quarry was the source of most of the stone quarried for church restoration work during the Victorian religious revival.

The 1873 plan (**5.6**) shows underground workings extending northwards from the northern face of the New Quarry. An annotation on the plan indicates that those marked in red date from

5.14. Underground workings in the Beer Stone at New Quarry, taken in August 1934. British Geological Survey photo P206334. Permit Number IPR/123-113CY British Geological Survey. © NERC. All rights reserved.

93

13 August 1884. However, there are further workings shown on the plan in rather indistinct pencil; these are marked right up to the edge of the sheet and clearly extended further north but could not be shown for lack of space. They are not dated, but are clearly later than the workings shown in red.

Stone is no longer extracted from the underground workings, but the opencast quarry was used until 2001 for the manufacture of lime by burning the Chalk which overlies the Beer Stone. Philippe Planel notes that the large plan of 1873 shows the different working methods used in the two quarries. In the area south of the lane, chalk for lime burning was quarried from above, and in two places there are annotations on the plan 'shot in and stowed' or just 'stowed', indicating that debris from chalk quarrying above was deliberately shot down in to worked-out areas of Beer Stone immediately below. On the north side of the lane, on the other hand, the quarry was pushed back as a continuous face, as a result revealing areas of pillars left over from earlier and deeper extraction of Beer Stone.

John Torrance writes that in the 1860s, competition from Bath and Portland, which had good railway connections, had almost killed off the Beer Stone business. However, Henry Ford, a local lime-kiln owner, saw the 1868 opening of the London and South West Railway branch line to Seaton as an opportunity to revitalise the freestone industry. In 1870 he took a 40-year quarrying lease of the hillside opposite the old Beer quarry, which included lime-kilns already owned by Ford. With six shareholders, he formed the Beer Freestone and Lime Co, the capital being £18,000. The idea was that the cost of developing the new stone quarry would be supported by the existing lime-burning enterprise, John Torrance quotes the brochure issued in 1882: 'A new quarry has been made, the superincumbent strata in this latter being burnt for lime. It is this fact…that enables the stone to come into the market at so reduced a price' [1/6d per cubic foot delivered to London, pro rata elsewhere].

Many papers referring mainly to the New Quarry workings, detailing especially the activities of the Beer Freestone and Lime Co, are held by John Scott of the Beer Quarry Caves. I am indebted to Philippe Planel for notes on these papers. They include: account books for the period 1840-1879; an 1890 plan of the New Quarry at a scale of 30ft to one inch, with detailed notes on the condition of the underground workings such as: 'set, crack, shivered roof, back, pillar split, overhanging roof, etc.'; an A5-sized map 'showing extension of quarry' on the north and 'Roman Quarry' on the south, with a section showing a narrow band of Beer Stone extending from Quarry Lane north through Bovey Lane; a file of correspondence relating to the Beer Freestone Company; a thick correspondence file with letters from Rolle land agents and replies from Henry Ford; an 1887 Prospectus of the Beer Freestone and Lime Company; and a Beer Stone Company booklet dating from about the turn of the century, with photos of men at work, underground crane, stone being loaded at Seaton Station etc.

A late Victorian photograph (photo 5.15) shows workings in the New Quarry. The remains of pillars of Beer Stone from earlier underground workings can be seen. Wooden cranes, horses and quarry workers can also be seen in the photo. A more detailed view is shown in 5.16.

John Torrance writes that by 1908 when Henry Ford died, the business was in trouble and was liquidated, only to be re-launched in 1911 under the name of Beer Stone Company by Henry Ford's two sons. The relaunch of the company, with a capital of £3,000, was recorded in an article of 21 November 1911 printed in *Pulman's Weekly News* (Clinton Devon Estates archive reference 22170). To celebrate the inauguration of the new company, a dinner was held in the Schoolroom, Beer, attended by 126 people including local worthies and the employees at the quarries. Various speeches were given; one by Mr W Henry Aplin reminded those present that about 30 years before, the late Henry Ford had held a similar dinner at the inauguration of the old Beer Freestone and Lime Company. Another speech by the Revd A H Hollis was on the history of the Beer Stone and its widespread use. He said that a sailor had once told him that the first thing that he saw when he arrived in Australia on one occasion was a pile of stone from Beer! The evening was enlivened with 'vocal duets and songs' and concluded with the singing of the National Anthem.

An advertising leaflet for the Beer Stone Company, dating from 1911, is shown in 5.17. It refers to the 'Home Depot, Seaton Station', where 'Thousands of Feet of Seasoned Blocks' were in stock. The Beer Stone Company was sold to Mr Perry, a local man, in about 1923. A photo (5.18) of 1926 shows roughly the same view as that in 5.15 but 30 years later. Contrast the busy scene of 1896 with this one, where there seems to be no sign of activity.

The extensive network of underground passages north of New Quarry was apparently used to store munitions during the Second

Above left: 5.15. A late Victorian (1896) picture of New Quarry, Beer showing the entrances to underground workings in the Beer Stone. Pillars of stone from earlier underground workings have been revealed as workings for lime have pushed back the face of the quarry. British Geological Survey photo P234142. Permit Number IPR/123-113CY British Geological Survey. © NERC. All rights reserved. *Above right: 5.16. A detail of the scene shown in photo 5.15 probably taken on the same day in 1896, showing men at work cutting Beer Stone.* British Geological Survey photo P234144. Permit Number IPR/123-113CY British Geological Survey. © NERC. All rights reserved.

World War.

John Torrance records that in 1939 the Beer Stone Company spent £4,000 modernising the quarry, and more kilns for lime-burning were being built. The lease was taken over in the 1950s by the Bath and Portland Cement Co; they closed the stone quarry, putting 19 stonemasons out of work, but the production of lime (which was then subsidised by the government), was kept going. Although the subsidy was stopped in 1976, the latest lessees of the quarry, Hanson Group, kept the lime-kilns going until 2001.

Before closure of the quarry in 2003, stone was extracted on a small scale on demand using large stone saws and used as dimension stone. A team of miners from the Bath stone mines was sent to the quarry and extracted about 500 cubic yards of Beer Stone which was stockpiled and reported to have satisfied demand for about ten years. Blocks of the stone are shown in the photo (5.19). An example of a

machine saw used in the New Quarry workings is on display in the museum just inside the entrance to Beer Quarry Caves.

New Quarry forms part of a Site of Special Scientific Interest (SSSI) designated especially for the geological interest of solution pipes filled with Clay-with-flints (described on page 64) that descend into the Chalk. The state of preservation of these features varies with the state of the quarry face and they cannot always be seen clearly. The 20th century lime-kilns are still to be seen in the New Quarry, with older ones in the hillside beyond.

The Hooken area

Henry De la Beche in his celebrated memoir of 1839 on the *Geology of Cornwall Devon and West Somerset* mentioned that 'A shaft was formerly sunk, and a level driven into the hill, between Beer Head and Branscombe, in order to obtain a continuation of the same [Beer

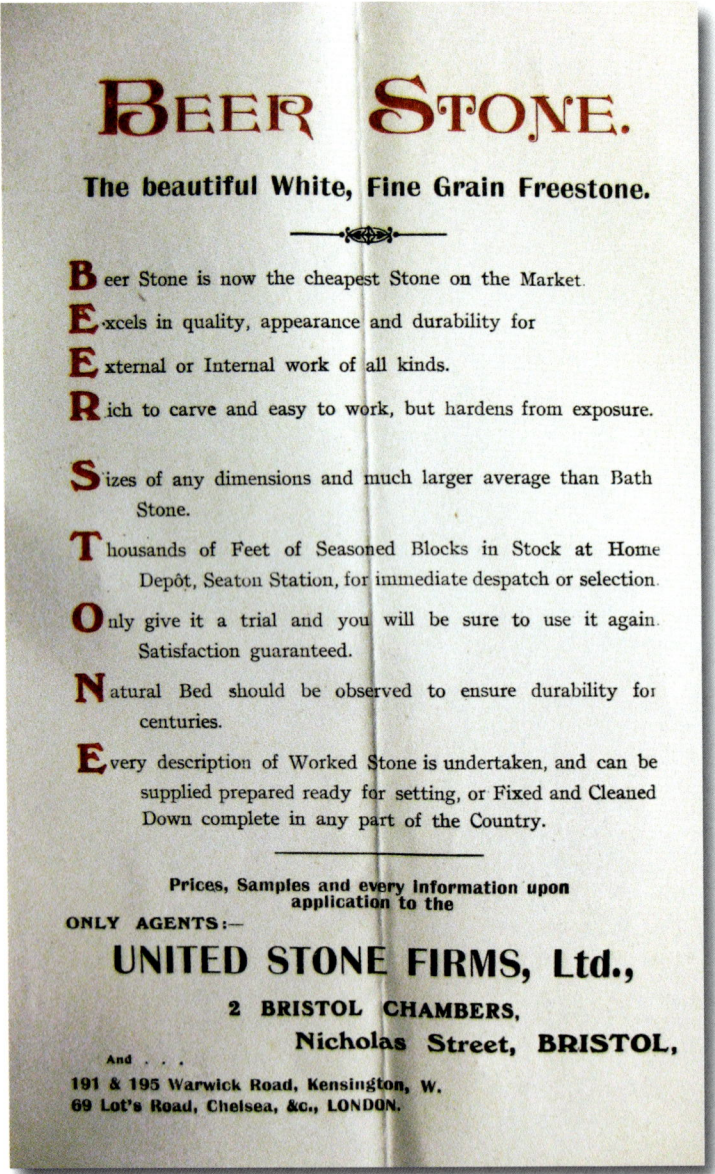

BEER STONE.

The beautiful White, Fine Grain Freestone.

Beer Stone is now the cheapest Stone on the Market.

Excels in quality, appearance and durability for

External or Internal work of all kinds.

Rich to carve and easy to work, but hardens from exposure.

Sizes of any dimensions and much larger average than Bath Stone.

Thousands of Feet of Seasoned Blocks in Stock at Home Depôt, Seaton Station, for immediate despatch or selection.

Only give it a trial and you will be sure to use it again. Satisfaction guaranteed.

Natural Bed should be observed to ensure durability for centuries.

Every description of Worked Stone is undertaken, and can be supplied prepared ready for setting, or Fixed and Cleaned Down complete in any part of the Country.

Prices, Samples and every Information upon application to the

ONLY AGENTS:—

UNITED STONE FIRMS, Ltd.,

2 BRISTOL CHAMBERS,
Nicholas Street, BRISTOL,

And . . .

191 & 195 Warwick Road, Kensington, W.
69 Lot's Road, Chelsea, &c., LONDON.

5.17. A leaflet, probably dating from 1911, advertising Beer Stone. Reproduced by courtesy of Clinton Devon Estates, reference 22170.

5.18. Another view of the New Quarry, taken in 1926 when there was no evidence of activity. British Geological Survey photo P249082. Permit Number IPR/123-113CY British Geological Survey. © NERC. All rights reserved.

5.19. Blocks of Beer Stone in the New Quarry, Beer, not long before closure.

Stone} beds, but the work does not appear to have been profitable, as it is now abandoned'.

Philippe Planel, examining Rolle estate leases, noted documents referring to Southdown and Hooken, not to Beer Quarry Caves, suggesting that these places were the site of workings. There is a problem with these references to workings at Hooken and Southdown, namely where were they situated? There are no surface quarries surviving in the area, but there is some evidence of underground workings in the past in the form of the adit in Hooken Cliff (5.3) and a shaft (see below). One possibility is that the workings were in an area which has since been carried away by landslip. Philippe Planel quotes Peter Youngman as suggesting that the Beer Stone may have been worked from the cliff face back inland before the Hooken landslip occurred in 1790. John Scott tells me that the Beer Stone adit was quarried from about 1650 and there was also a 45° slope shaft some distance inland (presumably the one noted by De la Beche in 1839). John reports that the entrance to this shaft was last seen in the 1960s when backfill subsided beneath the weight of farm machinery. Exploration of the adit at Hooken Cliff (pictured in 5.2 and 5.3), shows that it comes to a dead end after about 65ft, with no apparent connection to any workings farther inland.

A lease of 13 September 1681 (Devon Record Office 96 M/Box 34/10), granted by Edmund Walrond to Mathew Bragg, concerns existing and new quarries at Hooken. There is a mention of constructing 'one horse path from the common called South fields [?] to the said quarry ....to lay all such stones as hereafter shall be digged until they shall be carried away by sea or land'. Philippe Planel thinks that the reference to carrying away the stone by sea from Hooken is interesting since it reinforces the view that these are new workings since a suitable new path had to be built. Six late 18th century documents (1674-1691) relating to the Walrond family are all concerned with 300 acres of land at Hooken and Southdown and refer to freestone quarries as if they already existed (Devon Record Office 96M/Box 35/2).

Philippe Planel has examined Rolle freestone quarry leases dating from between 1797 and 1820 (Devon Record Office 96M/Box 33/9A). The 1797 lease granted to a Joseph Feltham of Hinton St George, Somerset, by John Rolle for 21 years is of particular interest. It gives the right 'to open one or more quarry or quarries for the commonly called Beerstone at a certain place called Hooken or Southdown'. The wording shows that the lease is not for the Beer Quarry Caves but suggests that this is a new quarry, especially since the wording 'dig search and delve' for stone occurs elsewhere in the lease. Note that this lease dates from after the Hooken landslip of 1790.

In the 1901 volume of the Geological Survey Memoir on the Cretaceous Rocks of Britain, dealing with the Lower and Middle Chalk of England, Alfred Jukes-Browne noted workings from the Hooken cliff adit, as follows: 'Dressed blocks of the stone still lie about on the slope below, alongside the pathway leading down to the shore, where the stone was shipped, but since the making of the railway to Seaton, the quarry has not been worked'. This interesting passage suggests that stone was extracted for use in construction of the branch railway line to Seaton in 1864-68.

HOW BEER STONE WAS WORKED

From the time of the Romans until about the middle of the 16th century, the stone was extracted entirely by hand using primitive hand tools such as picks. The sheer physical effort involved in patiently hacking away at the rock is remarkable. The initial opening, or 'breach', between the top of the Beer Stone and the overlying roof (the 'roof course' or 'cockly bed' of the miners) was made by pick in a space just large enough for a man to crouch. During the tour of the caves, the position of the breach can be identified on the upper part of the walls of the chambers since it is much rougher than the surface of the walls below where the stone was extracted. When saws came into use, the breach became narrower. The experience of the quarrymen enabled them to recognise beds of stone about 2 to 3ft thick which most people would not be able to distinguish (see illustration 5.7). They would drive in wedges at intervals along the bedding planes separating these beds thus splitting the blocks of stone away from the main mass of the rock.

From the middle of the 16th century, black powder was used to open the breach, and from that date the roof is much rougher in character. The amount of blasting powder used had to be judged carefully since too much would crack the Beer Stone below; it was used in combination with iron crowbars to lever out the loose stone from the breach. From about the early 18th century saws began to be used as well as picks and were used for pillar robbing. Pillars of stone were left by the earlier quarrymen to support the roof of the workings from collapse under the great pressure of the overlying rock. Their positioning and thickness were based on experience and intuition rather than any scientific measurement, and it was probably the case that they were in many cases unnecessarily thick for the loads which were present – a natural caution on the part of the earlier miners. Later quarrymen saw these large pillars as an easily exploited source of stone which could be had without the trouble of cutting from the main virgin faces of the mine – a process called pillar robbing. The miners judged that they could cut away stone while still leaving enough to support the roof – a fine judgement that fortunately seems to have worked since there is no obvious evidence of collapse following pillar robbing (5.20). Pillars near the entrance to the Norman workings were robbed first, and from there the 'robbers' worked their way deeper into the mine cutting away at the pillars. Many of the pillars show signatures and dates made by the workers (5.21); a mason might incise the pillar with his details to show pride in his status as a mason while a 'mere' quarryman would write with charcoal or lampblack.

The following description of the methods of working before saws were used is summarised from the fascinating account by John Scott and Gladys Gray in their booklet *Out of the Darkness*, by kind permission. The first operation was to open the breach, at the top of the Beer Stone. To get access to the top of the bed, 12 to 13ft above the floor, planks were pushed into slots cut into the Beer Stone face, forming a sort of scaffold. The quarryman climbed up this rough scaffold to the top of the bed, and crouching down, began hacking away at the rock with a pick. At first a pick with a 3ft-long handle was used, but as the breach got deeper, a pick with a longer reach was needed, so that successively longer handles were fitted on to the same pick head. The whole operation was intensely physical and carried out in a crouched contorted position (see 5.22) which brought the right shoulder, arm and face of the pick man in continual contact with the rough rock, causing grazing and burning

Top right: **5.20.** *Examples of 'pillar robbing'. Later quarrymen saw the large pillars left by earlier miners as an easily exploited source of stone which could be had without the trouble of cutting from the main virgin faces of the mine.*

Right: **5.21.** *Examples of writing and carving of masons' and quarrymens' details on pillars in Beer Quarry Caves.*

from the limy content of the rock. As John Scott and Gladys Gray have written so graphically: 'After only a few minutes' work he would be running with sweat which in the still, damp atmosphere, ran from his naked torso in rivulets. To stop work for a breather meant cooling rapidly and shivering with cold, so he worked to the very limits of his endurance'. It seems that under these conditions it is not surprising that the working life of a quarrymen was no more than 10 or 12 years.

After the initial breach was made, a man with a 'bedding axe' would take his place on the scaffold. His job was to smooth the bottom of the breach (i.e. the top of the Beer Stone bed) by swinging an axe with a head up to 12 lbs in weight. It was important to do this accurately so that regular-shaped blocks could be extracted.

The next operation was to cut vertical slots to isolate the selected block of stone, a process which involved cutting down the back and side of the block with a pick. The thickness of the subdivisions within the Beer Stone, shown on **5.7**, suggest that most blocks extracted would be between 2 and 3ft square, though some blocks

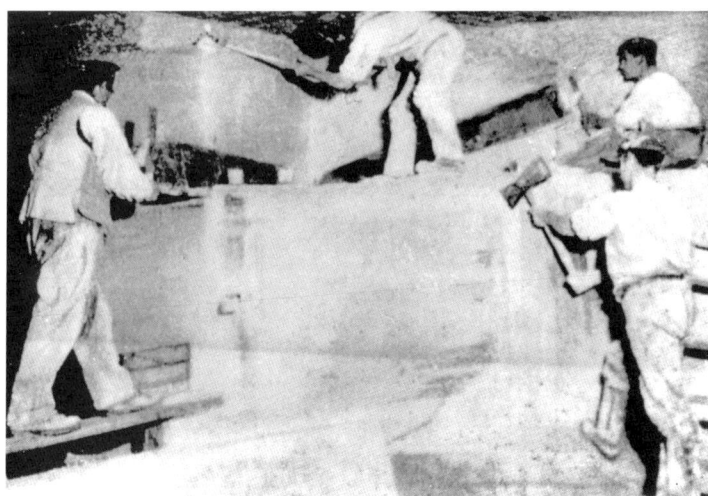

5.22. Quarrymen at work in the Beer Stone mines. The man at the top is opening the 'breach' at the top of the Beer Stone layer. Photo © Beer Quarry Caves.

Right: *5.23. A hand-operated crane in use lifting a block of Beer Stone in the Beer Stone mines.* Photo © Beer Quarry Caves.

were 4ft cubes (64 cubic feet) and such blocks would weigh about 4 tons. The largest recorded block was 4 x 8ft 6in x 12ft and weighed 26 tons. When the block was defined by the vertical cuts to rear and sides, it was freed from the face by driving in wedges at the base in careful sequence until the quarrymen were rewarded by the sound of a sharp crack which indicated that the block had broken away from the bed and was now free to move. Following the day's work, the block was tested by a tapper who struck it with a hammer. If it gave a clear sound it meant that the block was free of flaws; if not it might be rejected and the workers might not be paid.

Hand-operated cranes (5.23) were used to move stones within the mine. Individual blocks were lifted by inserting a device called a lewis into a hole drilled in the top of the block. This method could be used in confined spaces to lift stones, and also avoided damage to the finished stone by the use of chains slung round the stone. The lever arms of the lewis moved to grip the stone firmly when the weight was taken up. A lewis could also be used in the roof as an anchoring point for a crane.

A recurring unit of measurement for Beer Stone seems to have been a 'foote', and we have, for example, in a probate inventory on the death of William Walrond in 1720 the following entries (Devon Record Office 96M/Box 92/36): 'Six tunn and fourteen foote of free stone sold to Mrs Southcote £3 7s' and 'Six tunn twelve foote and a halfe sold to William Searle £3 6d'. There seems to have been a mixture of measurements based on weight and linear measurement.

As the cut blocks were removed they were measured and examined by a person called a 'touchstone' who assessed their value and calculated the payment due to the miners, depending on the quality of the work. It was thus in the miners' interest to make sure that the blocks were as square and neat as possible, a job which became easier with the introduction of stone saws. Each block, and subsequent smaller blocks cut from it, was marked with the position of the vertical since it was important to use the stone in the correct plane to avoid rain penetrating and causing the stone to deteriorate. John Scott and Gladys Gray remark on window mullions which contain perfectly cut sections adjacent to sections which have crumbled away because the mason had used the stone in the wrong plane.

The workmen toiled away in near-darkness, relieved only by the glimmer of tallow candles which were stuck to convenient places on the stone face using clay. It is said that the workers were allowed only five candles per man per day and consequently their use was carefully rationed. The fumes from the candles would make breathing difficult.

TRANSPORTING BEER STONE

The blocks were transported from the quarry by loading onto horse-drawn carts and in some cases carried considerable distances by this means. A photograph dating from about 1900 shows large blocks of Beer Stone at the mouth of the mine ready for onward transport (photo 5.24).

John Scott and Gladys Gray, in their guide booklet to the Caves, wrote that horses were always used, but until the 17th century oxen were also used, especially to pull heavy loads uphill. They noted that a primitive arrangement called a 'truckamuck' was in use until the 17th century. It consisted of a crude sledge made up of four small trees with branches roughly trimmed, lashed together. A horse was

5.24. Transporting blocks of Beer Stone from the New Quarry workings by horse and cart, about 1900. Photo © Beer Quarry Caves.

were some distance from the quarries; for example in Burlescombe Church which is 28 miles by road from Beer.

THE BEER STONE MINERS

Most of the miners probably hailed from Beer village, but many must have lived close to quarry, for example those involved in looking after the horses. The hard life of the miners has been described above. There were two divisions of labour: quarrymen and masons, and it is said that no man made the leap from quarryman to mason until 1856 when one William Cawley (whose name appears inscribed on a stone pillar – see photo **5.21** on page **98**) became a stone mason.

Other trades included carters and farriers and carpenters. Moving the blocks from the quarry to their destinations was a major undertaking. It is believed that over 400 horses were kept around the caves, and these needed farriers and other people to look after them. The large carts also needed maintenance by carpenters and others. In fact the scene on the ground about the caves must have been a hive of activity.

HOW BEER STONE WAS USED

Beer Stone is highly regarded by masons because it is easy to work when soft and hardens with exposure. It has an attractive creamy white colour which appealed to the medieval masons who prized its ability to give a sense of light within churches and cathedrals. It is easy to carve, and its fine grain size enables sharp detail to be carved quickly and easily, making it very suitable for internal details within churches, such as altar screens.

The high porosity (about 31%) and the tendency of the stone to absorb water in large amounts indicate that the stone has a limited resistance to weathering, and examples of the use of the stone on the outside of buildings bears this out; for example, the Beer Stone used in the image screen on the front of Exeter Cathedral (see photo **5.26**) has weathered badly in places, and extensive restoration work has had to be carried out (photo **5.27**). The results from Building Research Establishment tests of sodium sulphate crystallisation (indicating resistance to salt) suggest that the Beer Stone may be susceptible to damage from salt. However, the performance of the stone will be affected by the way in which the stone has been extracted, seasoned and laid in a building, and some stone may acquire remarkable toughness after weathering. It is generally

5.25. Beer harbour, from where much of the Beer Stone was loaded onto boats for coastwise transport to Exeter and other destinations.

harnessed to the tops of the trees and the roots dragged behind. If the stone was to be transported by water, it would have been carted to the coast at Beer harbour (photo **5.25**) where there must have been some mechanical arrangements for lifting the stone from the carts onto the boats. The known use of the stone in London buildings suggests coastwise water transport to the mouth of the Thames and into the heart of the capital.

W G Hoskins noted that in southwest England the stone was conveyed by water as far west as Tor Bay and the mouth of the River Dart, and less often farther west, for example in churches near the Kingsbridge estuary. However, even transport by water could be costly; Hoskins quoted an example where 32 cartloads of stone were bought by Exeter Cathedral in 1429-30, the cost at the quarry being 64s. However, the additional cost of transport from quarry to coast, ship to Topsham, and cart to the Cathedral, was £6 18s 8d. This illustrates that the stone was in such demand that it was worth spending considerable extra sums on transporting it to the desired location. Despite the cost of road transport the stone had such a good reputation that it was used in buildings, mostly churches, which

5.26. This photo shows how Beer Stone has been used in the image screen on the front of Exeter Cathedral.

5.27. Details of carvings in Beer Stone on the image screen on the front of Exeter Cathedral. That on the right is an original carving which has suffered badly from weathering over the centuries; that on the left is a modern but superbly carved replacement dating from 1982.

accepted that the Beer Stone is most suitable for inside work where high detail and sharpness of carving is required. Stone extracted from the New Quarry (formerly operated by the Hanson Bath and Portland Stone company and which closed in 2003) was used mostly in ecclesiastical work, whether for restoration or in the construction of new buildings.

Beer Stone has been used most widely in East Devon and Exeter as would be expected given the problems of transport of such a heavy and bulky material. However, the stone was so prized that it has also been used much farther afield, for example in famous London buildings such as Westminster Abbey, Westminster Hall, the Tower of London and London Bridge. In Beer village, there is, as might be expected, extensive use of Beer Stone, and in some places there is a striking use of a combination of Beer Stone and flint, for example in Diamond House, Fore Street, Beer (see photo **5.28**).

On a different scale altogether, Beer Stone has been turned into attractive art pieces, and many have been made by John Scott, proprietor of the Beer Quarry Caves. Pieces can sometimes be seen on display in the Marine House Gallery, Fore Street, Beer.

Exeter Cathedral

Perhaps the most interesting local use of the Beer Stone has been in Exeter Cathedral, where it is among at least eleven different types of stone used. Externally its most obvious use is in the image screen of carved figures on the west side (photo **5.26**). As I have mentioned, this screen illustrates the susceptibility to weathering of the stone when used outside, for many of the figures have suffered badly, and there has been extensive restoration. Photo **5.27** shows the contrast between restored and weathered carvings in the image screen. Beer Stone has also been used extensively inside the cathedral, and John Scott and Gladys Gray gave the following list of places where the stone can be found: parts of the walls of the nave; ribs of the vaulting and their smaller bosses; the chapels; the piscinae; the quire sedilla; all fine carving, moulding etc, with only some minor exceptions; the parclose screens; the screens of the Oldham and Speke chantries; probably also the pulpitum and Sylke chantries; and the Chapter House. In the Cathedral Close, some buildings contain Beer Stone removed from the Cathedral during periods of restoration.

Also in Exeter, the portico of Exeter Guildhall is made of Beer Stone, with pillars of granite.

Winchester Cathedral

Winchester and Exeter cathedrals liaised during the 14th and 15th centuries in the use of materials and masons. Beer Stone was one of the four main types of stone used. John Scott and Gladys Gray remarked that: 'At Winchester, Beer stone is probably seen at its most spectacular. Stand under the tower and look west along the Nave at the towering columns and graceful vaulting and almost everything to be seen is Beer stone, its creamy-whiteness giving light and warmth to the otherwise austere beauty. Over the High Altar the Great Screen is an outstanding example of carving in Beer stone...'

5.28. A combination of Beer Stone and flint used on the façade of the Diamond House, Fore St, Beer.

Beer Church, Colyton Church, Ottery St Mary Church, Cadhay House, Kentisbeare Church

These are other local buildings with interesting use of Beer Stone. In Beer Church, the exterior is of blue limestone (probably Jurassic in age) with mullions, quoins and porch of Beer Stone providing a pleasant contrast. The interior arcades are of Beer Stone, although the supporting columns are of Devonian 'marble'. The font and pulpit are of Beer Stone used in conjunction with Devonian 'marble'; the lectern too is of Beer Stone, carved by a local mason.

The main structure of Colyton Church is almost entirely of Beer Stone, as are many of the finely carved screens and monuments. The Saxon Cross in Colyton Church is of Beer Stone. In the small museum just inside the entrance to Beer Quarry Caves there is a large medieval window tracery carved from Beer Stone which was formerly part of Colyton Church, but which has been removed and re-assembled in the museum. Beer Stone is used throughout the fine church at Ottery St Mary, including its rib vaults, elaborately decorated ceiling, and in many screens and monuments. The exterior of Cadhay House, which is open to the public at certain times, is almost all Beer Stone. W G Hoskins, in his book on Devon, has commented on the striking use of Beer Stone in the tower of Kentisbeare Church, particularly in the newel staircase to the tower where it is used in a chequer affect with cinnamon-brown sandstone of Permian age.

BATS AT BEER QUARRY CAVES

Beer Quarry Caves are well known for their population of hibernating bats and are designated as a Special Area of Conservation (SAC) for the bats, and are also part of a Site of Special Scientific Interest (SSSI). David Wills considers that the caves have 'possibly the best assemblage of wintering bats in Devon'. Eight species of bat have been recorded, some of which are rare. Recording of the bats started in 1951, and in recent years showed the presence of about 70 to 80 individual bats belonging to an unusually wide range of species. These include Bechstein's bat (*Myotis bechsteini*) which is very rare, and the commoner Greater and Lesser Horseshoe bats (*Rhinolophus ferrumequinum* and *Rhinolophus hipposideros*). Other species recorded are Brandt's bat (*Myotis brandti*), Daubenton's bat (*Myotis daubentonii*), Natterer's bat (*Myotis nattereri*), the Brown Long-eared bat (*Plecotus auritus*) and the Whiskered bat (*Myotis mystacinus*). The presence of the very rare Bechstein's bat is the reason for the Caves' status as a European Special Area of Conservation. In 2006 an opening to a part of the workings that had been blocked for many years was re-opened and grilled for security to increase the area of workings available to the bats. Visitors taking the tour of the caves can sometimes catch sight of bats beginning their hibernation. The bats come in to the caves to begin their hibernation in October and leave in March to April. During a visit on 14 October 2009 I saw about half a dozen individuals which were probably Greater and Lesser Horseshoe bats, clinging to the upper part of the tunnel walls.

Chapter 6
BALL CLAY:
The Mines of North and South Devon

INTRODUCTION

About 65 million years ago, the period of earth history called the Tertiary began and great changes took place over our region. The Chalk oceans retreated, and southwest England became land during most of the Tertiary, with the nearest sea in the area of what is now the English Channel. Geologists divide the Tertiary into the Palaeogene below and the Neogene above, with the Palaeogene further subdivided into the Palaeocene, Eocene and Oligocene (see the table, 1.2). In north and south Devon during the later part of the Eocene and the Oligocene, clays, sands and lignites (a type of brown coal), were laid down by rivers and in lakes, and the clays include valuable 'ball clays' (described later in this chapter) which are widely used in the ceramic industry. These have been widely worked by various methods, including underground mining, which ended in 1999. Outside Devon, other Tertiary ball clays are found near Wareham in Dorset, where they have been dug for many years from mines and open pits.

DEVON IN TERTIARY TIMES 40 TO 25 MILLION YEARS AGO – LAKE BASINS AND RIVERS

During the late Eocene and Oligocene (about 40 to 25 million years ago) what is now southwest England was mostly a subdued plain across which rivers flowed, but a range of granite hills ran down its spine (6.1). The area was cut by several large faults, which are cracks in the earth's crust along which the rocks on either side have moved relative to each other. These faults, trending northwest-southeast,

had a major influence on the history of the area, because movements on them produced sags, or basins, in the earth's surface. There are several such basins in southwest England, the two largest and most important being the Bovey Basin in south Devon and the Petrockstowe Basin in north Devon. Another large basin (the Stanley Bank Basin) lies beneath the sea-bed east of Lundy Island. These three basins lie along the most important of the northwest-southeast faults, called the Sticklepath Fault after the village on the northern edge of Dartmoor. A small basin at Dutson near Launceston, on the Devon-Cornwall border, lies on a different fault line. The locations of the various basins and faults are shown on the map, 6.1.

Each subsiding basin acted as a giant settling trap for the vast quantities of clay, sand and organic material which poured into them from the surrounding areas. The basins were gradually filled up as they subsided, and considerable thicknesses of sediments built up in them. For example, a borehole proved the Petrockstowe Basin to be 2170ft deep, while the larger Bovey Basin has not been drilled to the bottom, but is estimated to be even deeper, possibly up to 3600ft. The late Eocene and Oligocene clay, sand and lignite deposits that fill the basins are collectively called the Bovey Formation, named after the town of Bovey Tracey at the northern end of the Bovey Basin.

The Bovey Basin

If you drive from Exeter towards Plymouth, you will cross an area of low ground between Chudleigh Knighton and Bickington. This apparently rather dull tract of country is underlain by one of the

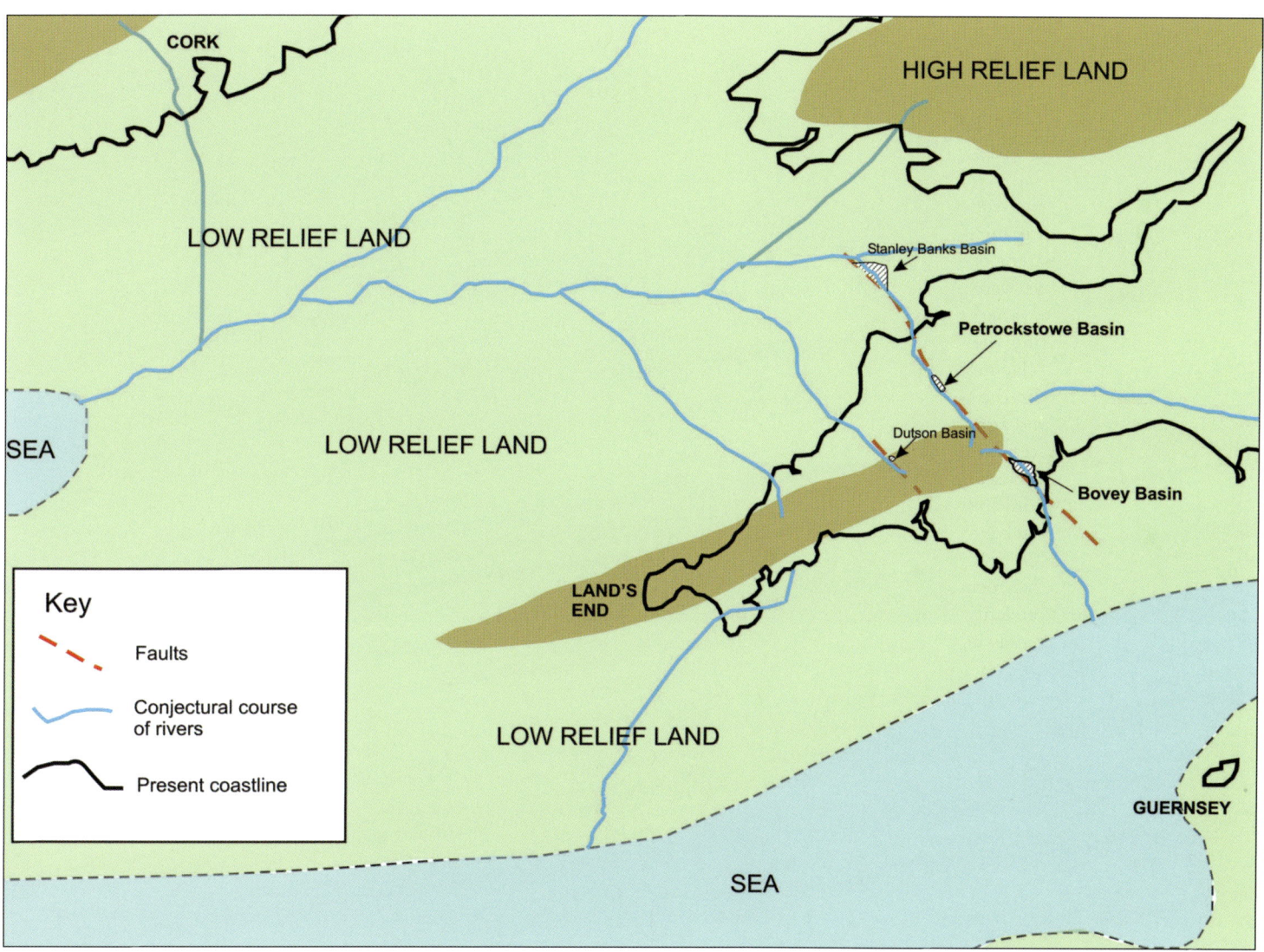

6.1. This map shows a reconstruction of the geography of southwest England during late Eocene to Mid Oligocene times (about 40 to 25 million years ago).
Map courtesy of Dr E C Freshney.

6.2. *This aerial view of the Bovey Basin, photographed in 2000 looking southwards, shows the extent of the large clay pits along the eastern side of the basin.* Photo courtesy of the Ball Clay Heritage Society.

most unusual and interesting geological formations in Britain. The area is the site of the Bovey Basin and is underlain by clays, sands and lignites of the Bovey Formation (map **6.3**). The deposit is best known for the 'ball clays' which are dug from large open pits for use especially in the ceramic industry. These pits can be seen in photo **6.2**, which is an aerial view looking southwards along the eastern side of the main Bovey Basin. This main part of the basin stretches for about 7 miles between Bovey Tracey in the northwest and Newton Abbot in the southeast and is about 4 miles wide. South of Newton Abbot there is also a smaller area of Bovey Formation deposits, which is called the Decoy Basin (apparently so named because wildfowl were decoyed there to supply winter larder for the occupiers of Forde House).

The Petrockstowe Basin

Like the Bovey Basin, the Petrockstowe Basin in north Devon (map **6.4**) formed along the line of the Sticklepath Fault. The basin is smaller and more elongate than the Bovey Basin, being about 4 miles long and about a mile wide. It is up to 2170ft deep, and is filled with a wide range of Bovey Formation sediments. The commonest types are fine-grained sand and very sandy clay, but, unlike the deposits of the Bovey Basin, lignite is generally rare. The sediments are coarser-grained and contain less lignite in the centre of the basin, but are finer-grained and more carbon-rich in two flanking areas, the northeastern and southwestern 'shelves', where the main ball clays occur.

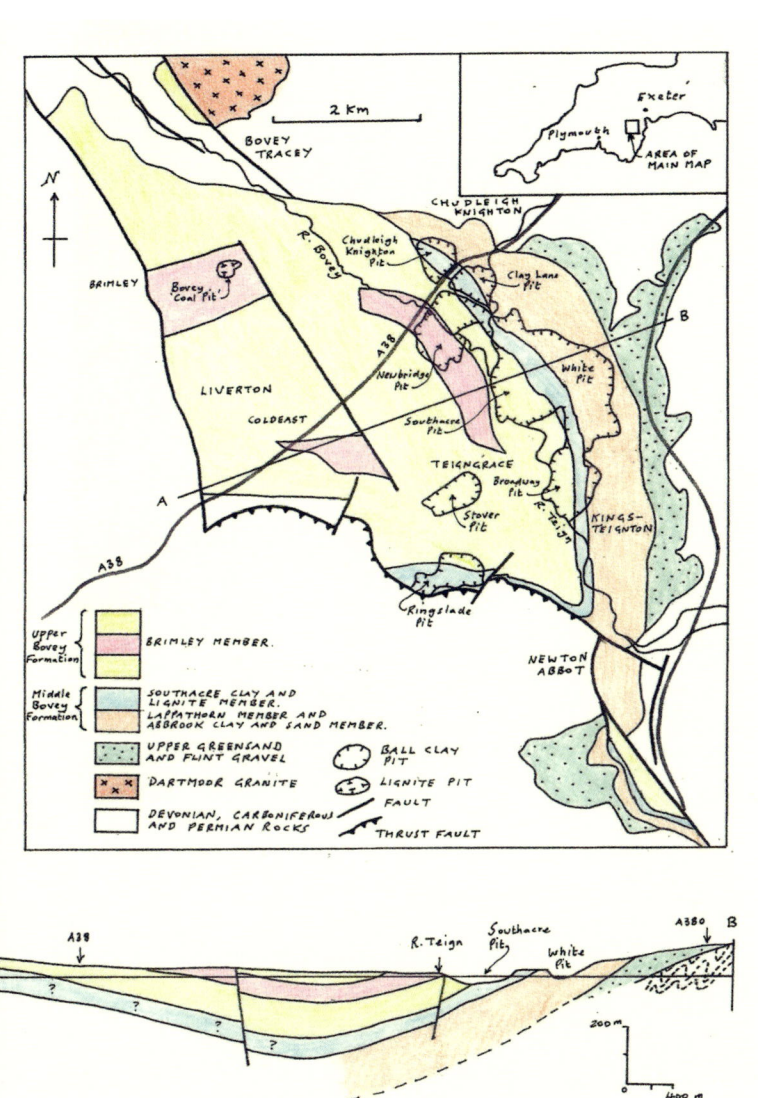

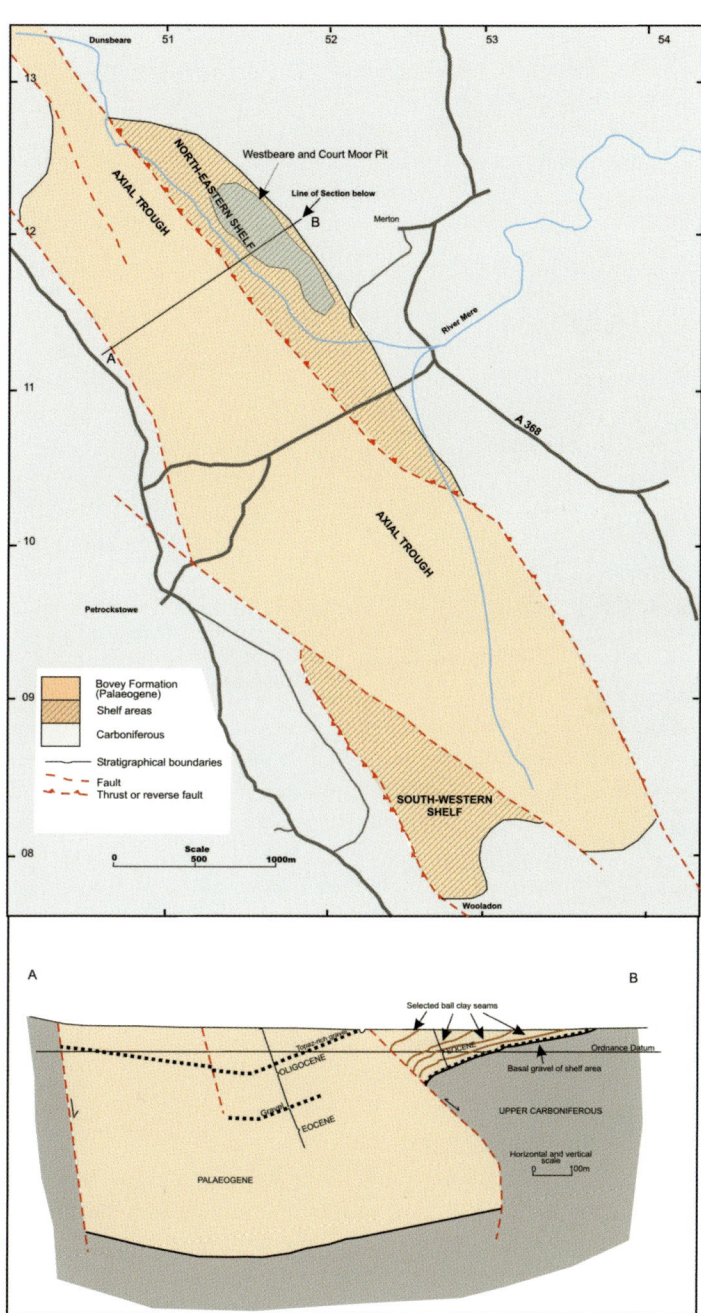

Above: *6.3. This map shows the simplified geology of the Bovey Basin, together with a cross-section through it. The geology is explained in the text.* Based on a map by Dr E C Freshney. **Right:** *6.4. This map shows the geology of the Petrockstowe Basin, together with a cross-section through it. The geology is explained in the text.* Map courtesy of Dr E C Freshney.

The climate in Tertiary times

Although there was a general trend for the climate to get cooler through the Tertiary, there was a time about 52 million years ago, during the Eocene, when it was subtropical. This period of humid subtropical climate was of great significance in geological and human terms, for it was then that the rocks of southwest England were subjected to intense chemical weathering, and deep weathering profiles were developed on a variety of rock types.

THE BOVEY FORMATION OF THE BOVEY BASIN

Good quality clays were first found and worked along the eastern side of the Bovey Basin between Chudleigh Knighton in the north and Newton Abbot in the south where, because of the westerly tilt of the strata, the main productive ball clay seams come to the surface (see the map, **6.3**). The main working area is still along this eastern side of the basin (see the panoramic view in photo **6.2**), although there are also large pits at Stover and on the southern edge of the basin at Ringslade.

The 1929 map (**6.5**) shows the two groups of commercial clays occurring along the eastern side of the basin and at Decoy, which had been recognised for many years. These consisted of a 'Stoneware Group' and an overlying 'Ball Clay' or 'Whiteware Group'. By the 1970s, the Bovey Formation along the eastern side of the Bovey Basin was divided by geologists into three 'members', in ascending order, the Lappathorn Member, Abbrook Clay and Sand and Southacre Clay and Lignite, which together comprise the 'middle Bovey Formation' (they are shown on the map, **6.3**). The last two correspond roughly to the older classification of Stoneware Group and Whiteware Group, respectively. The strata above the Southacre Clay and Lignite are included in the 'upper Bovey Formation' which is made up of seven members. Most of them consist of sand and clay, but two include much lignite. For simplicity only one, the Brimley Member, is shown on the map, **6.3**; it contains the lignites that have been worked around Bovey Tracey and which we will look at in the

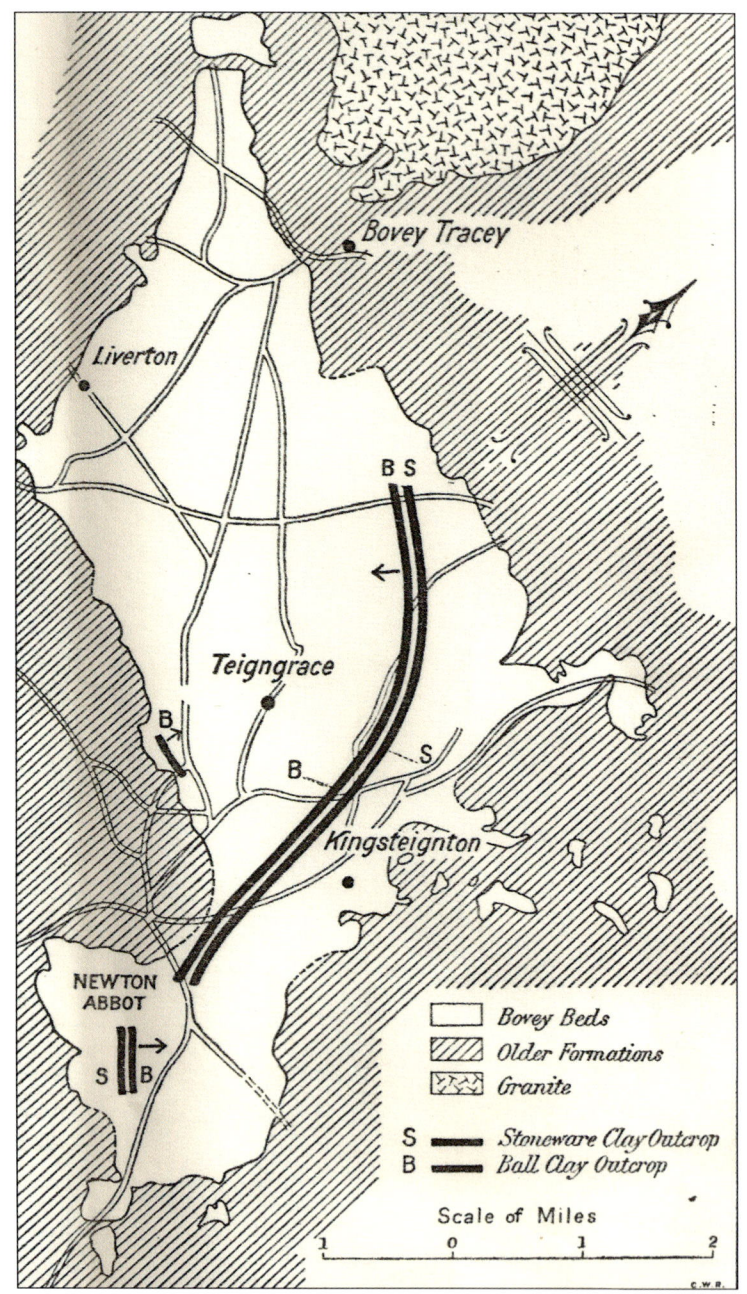

6.5. This map, reproduced from Scott's 1929 Ball Clay Memoir, *shows the division of the commercial clays of the Bovey Basin into a lower* Stoneware Group *(S) and an overlying* Ball Clay *or* Whiteware Group *(B). The general direction of inclination of the beds is indicated by the arrows.*

next chapter. Possibly up to 3000ft of 'lower Bovey Formation' lie hidden beneath later strata in the centre of the main Bovey Basin. Nothing is definitely known about these concealed strata because no deep borehole has been drilled to the bottom of the basin, but they may consist of sandy clays, silts, sands, and gravels.

THE BOVEY FORMATION OF THE PETROCKSTOWE BASIN

In contrast to the Bovey Basin, the Bovey Formation of the Petrockstowe Basin cannot be divided into distinctive subunits, or members. The strata are arranged in a series of 'cycles' in which the sediments are sandy or even gravelly at the base, but become finer upwards through sandy clays into, in places, smooth to slightly sandy clays, which may be very lignitic in the shelf areas. Below the topmost, usually carbonaceous, clay there are commonly rootlets which show that vegetation grew in place. In a small area in the northwest of the basin, there are finely laminated brown clays which contain the well preserved remains of fossil plant leaves.

HOW THE BOVEY FORMATION WAS DEPOSITED

The Bovey Basin

The sediments of the Bovey Basin were laid down in turbid shallow lakes, backswamps, and on 'deltas'. The gravelly muddy sands, found especially in the upper part of the Bovey Formation, were pushed out from rivers that flowed into the waters of a lake. Farther into the basin, clayey silts and clays were deposited in shallow lakes, and floating vegetation collected in very shallow backswamps farthest away from the basin entry, to form beds of lignite. Rootlets in a few clay beds indicate that some plants grew in place, but most of the lignite in the basin was washed in from dense redwood conifer (*Sequoia*) forests which clothed the uplands around the basin. The fossil plants found in the Bovey Formation of the Bovey Basin are mainly subtropical and tropical species, and include palms, ferns, heathers and many plants which grew in swamps. The plant fossils are of two kinds: those visible to the naked eye, such as leaves, stems, seeds etc. (examples are pictured in the next chapter, illustrations 7.5 and 7.6); and microscopic plant remains, mainly pollen grains. Studies of both kinds allow the Bovey Formation to be dated, and suggest that they are Middle to Lower Oligocene in age, with Eocene strata probably present deeper in the basin. There is however a possibility that the highest strata in the Bovey Basin are of Miocene age (page 144).

The Petrockstowe Basin

The sediments of the Petrockstowe Basin were mainly laid down by meandering rivers rather than in a lake. However, well-laminated clays in the northwest of the basin suggest that there a temporary lake may have formed. In the central part of the basin the sediments formed along river channels, while in the shelf areas they were mainly laid down on river floodplains and in swamps. Thicker beds of gravel in the centre of the basin may have been deposited by more energetic braided rivers. The river which deposited most of the sediments in the Petrockstowe Basin flowed through an area of relatively low relief in subtropical conditions, with a plant cover that included conifers on the higher ground, together with ferns and swamp-loving plants near the river and feeder streams. Minerals in the sands and gravels originated from the Dartmoor Granite, proving that the river flowed from the southeast towards a sea in the outer Bristol Channel and Celtic Sea.

No animal fossils have been found, probably because any bones would have been dissolved in the acid waters, but it is inconceivable that mammals, birds and insects did not abound. Studies of the fossil pollen in the clays indicate that most of the Bovey Formation in the Petrockstowe Basin is of Eocene age, with Oligocene beds present in the uppermost 800ft or so in the central part of the basin (this is shown on the cross section in 6.4) The strata in the shelf areas of the basin are entirely Eocene in age.

BALL CLAYS

Like china clay, ball clay contains the clay mineral kaolinite, but the two types of clay formed in quite different ways. China clay, best known from the St Austell area of Cornwall, but also found in south Devon near Plymouth, is produced by the hydrothermal alteration and subsequent weathering of granite in place. It is only after it has been washed out by the efforts of man from the parent granite, using powerful jets of water, and is allowed to settle that it becomes a true clay. On the other hand, ball clays are sedimentary, that is, they have been transported naturally by rivers and streams and then deposited away from their original place of origin.

Ball clays are highly plastic, rich in the clay mineral kaolinite, and fire to a white or near-white colour in the oxidising atmosphere of a kiln. The term 'ball clay' originates from the earliest methods of hand working in open pits. The clays were cut into blocks about 9 inches cube, and these acquired rounded edges during handling and transport, giving them a ball-like form. The methods of working involved the use of specialised tools – see page 118 below and photo 6.9.

For many years it was thought that ball clays were derived entirely from weathered granites, from which the clay portion was washed out into the adjacent basins. However, more recent research has shown that the clays were derived from chemical weathering profiles developed on a variety of parent rock types by intensive tropical weathering during the Tertiary. In fact, in the case of the Petrockstowe Basin, the clay minerals making up the ball clays were derived almost entirely by the weathering of shales and sandstones of Carboniferous age found around the basin. For the Bovey Basin, a much wider range of materials from around the basin contributed to the ball clays. In some clay seams, the material was derived mostly from the Dartmoor Granite and thus these clays are very like china clay - that is, much coarser grained and thus less plastic than ball clay, but firing white. Other clays were derived from weathering profiles developed on other rocks and these tend to be finer-grained and thus more plastic, and firing off-white to creamy-white. The more varied source materials are the reason why such a variety of clays is known from the Bovey Basin, and as many as 270 different types have been recorded, apparently the widest range known from any ball clay deposit.

The deep tropical weathering was particularly important for the economic value of the ball clays. Most clay deposits in the world are coloured brown, red or yellow owing to the presence of iron oxides, so that they do not fire white in kilns and are therefore only suitable for making items such bricks, tiles and pipes. On the other hand, the intense weathering of the Bovey deposits and their source rocks removed most of the iron oxides so that the resulting clays are suitable for producing white wares.

Ball clays are mainly made up of mixtures of three minerals: kaolinite, mica and quartz. The variations in the amounts of these minerals, plus the presence of carbon, provide the wide range of ball clay types mentioned. Clays with significant amounts of finely divided lignite are often black or brown, plastic, and white-firing; they are traditionally called 'Whiteware' clays or 'Potters' clays. Less plastic clays with little carbon but appreciable amounts of fine-grained quartz sand are called 'Stoneware' clays. The ball clays of the Bovey Basin have been divided into four groups, each of which has distinctive properties, summarised as follows: *Group 1 Clays ('Black*

6.6. Southacre Clay and Lignite in Newton Abbot Clay Co Pit, April 1970. It consists of alternating layers of black lignite and brown clay. At that date the clays were cut using pneumatic spades.

Ball Clays'), found particularly in the Southacre Clay and Lignite, consist of brown carbonaceous clays interbedded with lignites (photo 6.6). The clays are particularly noted for their exceptional whiteness when fired. *Group 2 Clays ('Dark Blue Ball Clays')* are found in the lower part of the Southacre sequence and are probably the strongest clays in the Bovey Formation. They fire white to off-white. *Group 3 clays ('Light Blue Ball Clays')* occur in the higher part of the Abbrook Clay and Sand Member. They are generally low in carbonaceous matter and fire white to off-white and ivory. They are particularly useful in the liquid slips used by the manufacturers of cast sanitaryware. *Group 4 Clays ('Stoneware Clays')* occur in the Abbrook Clay and Sand and are siliceous ball clays, firing off-white to buff. They contain little or no carbon and have high percentages of free quartz. They are the coarsest grained ball clays and are used in stoneware manufacture and as refractory bonding clays.

The ball clays of the Petrockstowe Basin differ somewhat from the Bovey Basin clays, and the four-fold grouping of the Bovey Basin ball clays cannot be applied. Broadly speaking, there are two distinct clay types in north Devon. Both are made up mainly of the clay mineral kaolinite, but one of the types also contains appreciable amounts of the clay mineral montmorillonite and this is partly the reason for the high strengths and plasticity of the north Devon clays. The thick ball clays tended to form on flood plains and in backswamps on the flank or shelf areas of the basin (shown on the map, **6.4**). The quarries at Westbeare and Courtmoor (map **6.4**) were developed by the ball clay company Watts Blake Bearne to produce different types of clay: at Courtmoor the '120s' were suited for blends of clay used in sanitaryware and were taken to Newton Abbot for that purpose. Clays from Westbeare – the '110s' and '116s' - were used in blends developed by the North Devon Clay Co.

CLAY NAMES

Over the years a fascinating range of local descriptive terms for ball clays has evolved, based on colour (both fired and unfired), usage of the clay, and physical characteristics. Descriptions in quotes in the list below are from the glossary to Alex Scott's 1929 *Ball Clay Memoir*. Many of the names refer to the colours of ball clays and we have for example, Black Ball Clay, Dark Ball Clay, Blue Ball Clay, Light Blue Ball Clay, Ivory Ball Clay, Grey Ball Clay, White Ball Clay etc.

Alum – 'Term applied to clays high in alumina'.

Brokes – 'Clay which will not cut into "balls": is generally very short [see below] and has bad colour on firing'.

China-ball clay – 'Term applied to a light coloured, highly aluminous, white-burning clay'.

Corny clay – Clay containing 'corns' which are 'Grains of sand, quartz or hard clay ranging up to 0.5cm in diameter'. The grains can also be made up of the mineral sphaerosiderite.

Cutty clay – 'Plastic clay used for the manufacture of tobacco pipes'.

Figgy clay – 'Clays with a finely nodular or roughened appearance, like that of figs'. The nodules are made up of lignite.

Glady or gladii clay – 'Variegated black and white clay often associated with stoneware clays, but used for the manufacture of white ware'.

Hawse or hawsy clay – ' "Crumbly" clay which becomes plastic when worked up; sometimes "auze" or "auzey"; supposed to be derived from "rose"'.

Household clay – 'White relatively non-plastic clay used for making blocks for whitening stone steps, sills, etc.'

Lice – 'Very thin beds of sand, lignite, etc.'

Mottled clay – 'Clay which in raw state is variegated – yellow, pink, purplish, brown, etc. – used for stoneware manufacture'.

Pinks – 'Stoneware clay of pink colour'.

Pipe clay – 'Clay used for the manufacture of tobacco pipes'.

Potters clay or Potting clay - clay suitable for the manufacture of pottery.

Saggar clay – A crude variety of refractory clay used for making the 'saggars' in which pottery is placed in the green state before biscuit firing (L T C Rolt, 1974, page 34).

Short clay – Non-plastic clay.

Stilt clay – 'Vitreous clay suitable for the manufacture of stilts'.

Stoneware clay – Clay suitable for the manufacture of stoneware.

Top or Coarse Top – 'Stoneware clay which is used with admixture of other clay'.

Whiteware clay – White-firing clay suitable for the manufacture of pottery.

There are many combinations of terms, for example 'Best White Potting Clay', 'Black Corny Clay', 'Black Hawse Clay', 'Potter's Black Ball Clay', 'Short Pinks', etc.

THE HISTORY OF BALL CLAY WORKING

I am indebted for the content of parts of this section to the Ball Clay Heritage Society's account in their book *The Ball Clays of Devon and Dorset*. As with so many enterprises, the Romans are reputed to have been the first to exploit the ball clays of the Bovey Basin, but there seems to be no direct evidence of this. L T C Rolt wrote that 'the beds lay undisturbed until the beginning of the 17th century', but it seems unlikely that clay digging of some kind for the production of crude pottery did not continue at least sporadically throughout this long period.

The ball clay industry is generally believed to have been stimulated by Sir Walter Raleigh's introduction of tobacco into England in the 16th century and the resulting demand for tobacco pipes – hence the early use of terms such as 'pipe clay' and 'cutty clay' ('cutty' referring to a short-stemmed type of tobacco pipe). However, the very plastic ball clays, when used on their own, were not particularly suitable for tablewares, owing to the fact that they contracted and expanded in the kiln when fired, making the process difficult to control. Thus in the 17th century it was common for tablewares to be made of coloured local clays, covered either with a thick white glaze (as for example the tin-glazed Delftwares), or by a slip of white clay. In Bideford, for example, potters used white clay slip from the Petrockstowe Basin through which designs were scratched to expose the coloured clay body underneath.

Evidence of a flourishing early trade in pipe clay is provided by a 1662 Act of Parliament, not repealed until as late as 1853, which forbade the export of such clays abroad. In the 17th and 18th centuries, clay from north Devon was exported through the port of Bideford, destined especially for Bristol pipe makers, the first recorded shipments being in the 1650s. There is a record of small quantities of clay being shipped from Bideford to the Port of Chester in 1691, and by 1730 the amount shipped had trebled. In contrast, shipments of clay from south Devon seem to have been of limited extent until the mid-18th century.

European potters struggled to match the superb white porcelain that had been made in China for centuries, and it was only in the 18th century that they succeeded in producing good-quality white pottery. In England, celebrated early Staffordshire potters, such as Josiah Wedgwood, learnt how to produce successful combinations of plastic ball clay, relatively non-plastic china clay, and other ingredients. These recipes required ball clay from Devon (and Dorset) together with china clay from Cornwall and Devon, thus providing an intensive stimulus for the industry. The export of ball clays increased rapidly from 500 tons in 1740, to 2,381 in 1765, 4,069 in 1770, and 9,195 in 1785. Two-thirds of these amounts went to ports in the Mersey estuary and the remainder went to ports ranging from Newcastle to Swansea. There was also an important international trade, especially to Europe and the USA, by the end of the 19th century.

Richard Polwhele in his 1797 *History of Devonshire,* described 'pipe' and 'potters clay' at Kingsteignton, and noted that 'a considerable quantity is annually sent from the harbour of Teignmouth to London, Liverpool, and all the northern potteries, to the amount of at least ten or twelve thousand tons'. He also recorded that in north Devon '...in the valley between Merton and Petrockstow, are dug large quantities of fine white pipe-clay.' He gave a description of working the clays using traditional tools to cut the clay into 'balls'. In 1808 Charles Vancouver wrote that 'Large portions of pipe clay were formerly dug on the demesne land of Petrockstow, and shipped round from Bideford to the potteries in Staffordshire'.

In the account in the Lysons' *Magna Britannia* of 1822, based largely on earlier accounts by William Maton and Charles Vancouver, the annual amount of ball clay exported from Teignmouth in about 1820 was recorded as 20,000 tons. The high cost of packhorse or horse and cart transport from the Marland area of the Petrockstowe Basin to Bideford seem to have led to closure of the works in the early 19th century; Lysons noted that clay pits in Petersmarland and Petrockstow had not been worked for 20 years prior to 1822, and this may have been because of competition from workings in the Bovey Basin which had the advantage of better communications.

The repeal of the ban on ball clay exports in 1853 resulted in increasing exports to a wide variety of markets in Europe and North America. In England, Staffordshire remained the centre of pottery-making despite its distance from its clay resources, mainly because of the expertise and traditions that had been built up there, and the availability of cheap coal from the adjacent coalfields for firing kilns. There were however some areas outside Staffordshire, conveniently placed for access to supplies of ball clay, that developed a thriving pottery industry, for example the Poole Pottery in Dorset, and of more immediate interest to our story, the various potteries in Bovey

Tracey. Photo 7.2 shows the surviving bottle kilns of the former Bovey Pottery Co Ltd (now the site of the House of Marbles tourist attraction). It is also interesting to note that before the availability of cheap coal, the kilns of the potteries were fuelled by lignite from the nearby Bovey 'Coal Pit' (see Chapter 7).

In both the Bovey Basin and Petrockstowe Basin the stoneware ball clays were used for the manufacture of tiles, bricks, drain-pipes, chimney pots, etc. Pale cream bricks from these sources are common in buildings in some Devon towns, for example Kingsteignton. The hard pale yellow bricks from Marland were widely used in north Devon and can be seen in buildings in late Victorian resorts such as Ilfracombe. One of the most notable companies producing stoneware products was Candy & Co (now British Ceramic Tile) located at Heathfield in the centre of the Bovey Basin: it even produced some interesting art pottery for a while. Other companies producing similar products were Hexter, Humpherson & Co in Newton Abbot and the Marland Brick & Tile Works in north Devon. As well as conventional bricks the Marland works produced paving and fire bricks and were also well known for their wide range of architectural items in terra cotta.

The English ball clay industry suffered from the effects of two World Wars and the intervening slump, as well as the opening up of ball clay deposits in north America, so that by 1945 annual production was only 90,000 tons. The industry was also slow in adopting new technology, and a 1946 Board Of Trade Inquiry noted that '....it would be difficult to find any industry in this country where there has been so marked unawareness and such lack of initiative on the part of many producers to modern industrial change'!

However, after the Second World War the industry in Devon and Dorset enjoyed sustained growth; by the 1950s improved standards of living in Britain and Europe led to strong demand for sanitaryware and tiles. There was a very marked improvement in the adoption of new methods, to the extent that by 1969 the firm of Watts Blake Bearne & Co. Ltd was awarded the Queen's Award for technical innovation in underground mining. During the 1960s and 1970s there was strong growth in Italian and Spanish ceramic production. In the 1970s and 1980s the Middle and Far East emerged as valuable new markets, and in the 1990s there was further growth in ceramic tile production. By 1970 annual UK sales of ball clays had grown to 700,000 tonnes, and in 2000 reached a record 1.1

million tons. Since 2005 annual sales have been around the million ton mark. The value of UK ball clay production in 2008 was about £82 million.

Today the UK remains a major world producer and exporter of high quality ball clay and Devon is the most important source in the UK with 75% of total sales in 2010. Sales from the Bovey Basin were 500,000 tonnes and from the Petrockstowe Basin 255,000 tonnes. Sales of ball clay from the Wareham Basin in Dorset were 250,000 tonnes. Domestic sales of ball clay have declined, but there has been a steady growth in export markets. Exports of sanitaryware clays are particularly important. Exports of UK ball clays were 853,000 tonnes (83% of total sales) in 2008.

BALL CLAY COMPANIES

The detailed history of the many different companies that have worked the ball clay deposits of Devon is outside the scope of this book, but I have included a short summary here. More detail can be found in L T C Rolt's 1974 book *The Potters' Field*. There is also a very useful description in the Ball Clay Heritage Society's 2003 book *The Ball Clays of Devon and Dorset*. Michael Messenger's 2007 book *North Devon Clay* is an invaluable source for north Devon.

The Bovey Basin

For much of the history of ball clay working, land in the Bovey Basin was in the hands of great landowners such as the Bishop of Salisbury and the Cliffords of Ugbrooke (around Kingsteignton), the Earls of Devon (around Newton Abbot) and the Templers and the Dukes of Somerset (around Stover). These families granted mining leases to clay merchants and were not directly involved in the industry. Canals were financed and built by the large landowners, for example the Stover Canal by James Templer II and the Hackney Canal by Lord Clifford. Some clay merchants gained the freehold ownership of important clay-bearing land from the large landowners, sometimes as well as having long-term mining leases. For example, the Watts family bought key clay lands and a lease from the Bishop of Salisbury, which in 1796 they brought into a partnership with Whiteway. In 1856 the Devon and Courtenay Clay Company was formed with a lease of the Decoy property from the Earl of Devon, thus creating competition for Whiteway, Watts & Co. In 1861 the Watts family broke away from the Whiteways and formed a partnership with Blake and Bearne, forming Watts, Blake, Bearne and Company. Whiteway, Watts & Co then became simply Whiteway and Company.

After the First World War, a number of new companies were set up or became involved in the ball clay industry. In the Bovey Basin, these were Hexter and Budge Ltd, the Newton Abbot Clays Ltd, the Mainbow Clay Company Ltd, the Pochin Ball Clay Company Ltd and the London, Australian & General Exploration Company Ltd. In 1929 a snapshot of eight companies working in the Bovey Basin was given by Alex Scott: Messrs. Whiteway Wilkinson, Messrs. Watts Blake and Bearne, Messrs. Hexter and Budge, Pochin Ball Clay Co Ltd, Newton Abbot Clay Co, The Ringslade Clay Co, Messrs Candy & Co Ltd, and Devon and Courtenay Clay Co Ltd.

Following the Second World War, about fifteen companies were working ball clays in the Bovey Basin, but by 1969 as a result of mergers or takeovers there were only two major producers: Watts Blake Bearne & Co Ltd and English China Clays. In 1981 the two companies operating in the basin were Watts Blake Bearne and Co Plc (WBB), and ECC Ball Clays Ltd (formerly Hexter and Budge (ECC) Ltd). Ball clays were at that time being extracted from thirteen open pits and seven modern adit mines.

During the Second World War the Pochin Ball Clay Company and the Mainbow Clay Company had joined what became English China Clays (ECC). In the 1950s ECC acquired the London, Australian company and Hexter and Budge. During the 1960s there was competition between WBB and ECC to take over other companies. For example ECC acquired Meeth (North Devon) and Pike Brothers, Fayle (a merger of the two Dorset companies), operating at first as Hexter and Budge and then as ECC Ball Clays. After these acquisitions, ECC had a share of about 40% of UK ball clay production. Today, the parent company, English China Clays Plc, is owned by a French company, Imerys SA and is called Imerys Minerals Ltd. In the 1960s, under the chairmanship of Claude Pike, WBB became a public company, and acquired Devon and Courtenay (which had itself acquired Whiteway & Co), Newton Abbot Clays and (in 1967) the North Devon Clay Company.

Since 1999 WBB has been wholly owned by SCR Sibelco SA of Belgium. In 2001 WBB was merged with Sibelco Minerals & Chemicals, the former Hepworths sand division based in Cheshire, to form WBB Minerals. WBB's headquarters and laboratories in Newton Abbot were closed. In 2008 WBB Minerals was renamed Sibelco UK. In 2010 there were two large companies working in the

Bovey Basin: Sibelco UK and Imerys Minerals. Sibelco is the largest producer of ball clay not only in the UK but in the world, and operates in south and north Devon. Imerys had interests in both the Devon basins and Dorset, but ceased production in the Petrockstowe Basin in 2004.

The Petrockstowe Basin

In contrast to the numerous changes of ownership of ball clay firms on the Bovey Basin, ownership in the Petrockstowe Basin was for many years in the hands of just one company, the North Devon Clay Co. As a result, working practices stayed largely unchanged in north Devon, while the frequent changes of management in the south led to more technological innovation, to the extent that the south Devon firm of WBB eventually took over the North Devon Clay Co.

As long ago as the 1650s, the Greening family, merchants based at Bideford, owned part of what was to become the Marland clay works. William Wren, a descendant of Greening, established the Marland Brick & Clay Works Limited in 1879 and built the vital Torrington & Marland light railway, opened in 1881, from the works to Torrington (see page 134). The North Devon Clay Company was mentioned in entries in *Kelly's Directory* in 1881 and again in 1889. It was not a limited company, but its existence indicates that the business of clay extraction on the one hand and brick-making on the other had become separate by this time. The Marland Brick & Clay Works Limited company lost money each year and was wound up in 1888 after only eight years. In September 1891 a new company, the Marland North Devon Brick Company Limited, was formed. The clay pits seem to have been operated by the North Devon Clay Company, which dug the ball clay for sale to the brickworks as well as other markets.

Lord Clinton leased land beyond Clay Moor (the site of the brickworks) southwards towards Bury Moor to the North Devon Clay Company. With this expansion in the business a limited company was formed, the North Devon Clay Company Limited, incorporated on 20 May 1893. The Marland North Devon Brick Company was bought out and the lease surrendered to the North Devon Clay Company Limited in January 1894. Thus the whole of the clay pits, brickworks and railway were now controlled by the North Devon company. In 1967, Watts Blake Bearne & Co Ltd, the south Devon clay producer, took a large stake in the share capital of the North Devon Clay Company Limited, and in 1968 the north Devon company became wholly owned by the south Devon giant. As a result, WBB became the largest UK ball clay producer, with a share of about 60% of the total market. Subsequent changes in ownership are noted above, on page **116**.

In 1920 the Meeth (North Devon) Clay Company Ltd was set up to work the deposits in the southern part of the Petrockstowe Basin near Meeth. The share capital was bought in 1965 by English China Clays Limited, and in 1967 the assets were transferred to a subsidiary of ECC, Hexter & Budge Ltd. The name of the Meeth (North Devon) Clay Company was changed to Meeth Ball Clays (ECC) Limited, which was wound up in 1979. In 1970 Hexter & Budge Limited was renamed ECC Ball Clays Limited. The French company Imerys took over the ECC Group in 1999.

THE BALL CLAY WORKINGS

The Bovey Basin

The ways in which ball clays have been won from the ground have evolved over the many years that they have been exploited. The diagram (6.7) summarises the various methods of working in relation to the various parts of the dipping ball clay sequence. In the earliest days, from the 17th century, the clays were probably dug from what were virtually just shallow trenches which barely penetrated much below the gravelly overburden ('head' or 'ridding') that is widespread on top of the clay beds. Their depth was limited by the problem of water inflow which continually flooded the pits. However, with the development of pumps in the 19th century, there was a natural increase in the size of pits to what may be described as small open pits which were larger and deeper than the trenches that preceded them. In the Bovey Basin shallow open pits were deepened by using the 'square pit' technique which meant that it was possible to work deeper seams of clay.

From the second half of the 19th century, shafts were sunk to the clay seams at depth, and levels driven out from the base of the shaft to work the clay. A later development of underground mining was the adit method, first used in the 1930s, which in the 1960s became the main method replacing shaft mining until all underground mining ceased in the Bovey Basin in 1999. It involved driving an inclined tunnel from the outcrop of the seam down the dip.

From the earliest days ball clays have been, and continue to be, worked in many open pits, but large-scale open pits began to dominate production from the 1950s. Even in the late 1960s the pits were of a fairly modest size, and were worked mainly by hand using pneumatic face shovels (see photo **6.6** dating from 1970), but today, the pits are very large, and previously separate pits have been amalgamated into 'super-pits', with consequent economies of scale and ease of working using hydraulic excavators and dump trucks.

6.7. This diagram shows the westward-dipping ball clay sequence along the eastern side of the Bovey Basin and the various ways in which it has been worked. The 'Lignitic Sequence' corresponds roughly to the Southacre Clay and Lignite Member of map 6.3. Illustration courtesy of the Ball Clay Heritage Society.

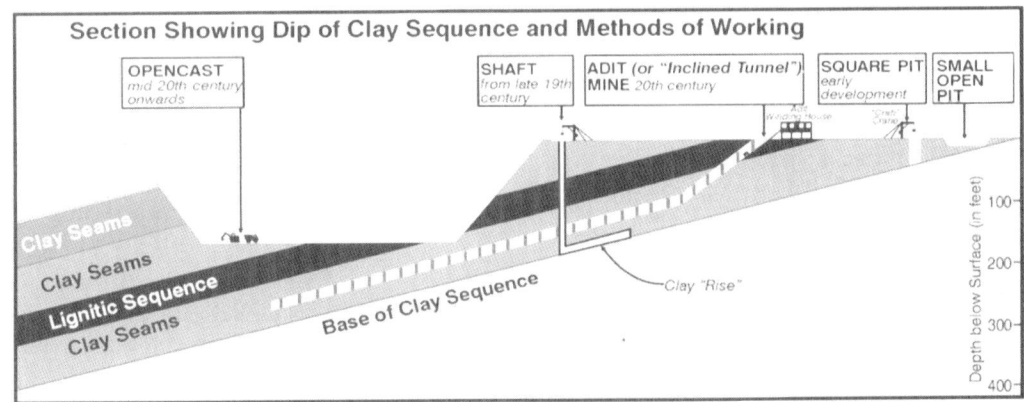

117

Small open pits

In the working of small open pits, the overburden was first removed until the clay seam was revealed, which the 'claycutters' then dug out by hand. A clean floor of clay exposed in the base of a pit was cut across at regular intervals about 9 inches apart, a process called 'long-scoring'. It was then cut across again at right angles (a technique called 'thirting' or 'thwirting') to give a criss-cross pattern of squares about 9 inches across. The cutting was carried out using 'thirting' (or 'thwirting') irons which were heavy iron spades with blades about 4 inches wide. The clay blocks were then undercut and levered out using a mattock-like tool called a 'lumper'. A less heavy version of the lumper, called a 'tubil' or 'tubal', was used to cut clay underground and to trim the workings. The cube of clay released by the 'lumper' weighed about 36lb. The cubes were then stabbed with a curved iron spike called a 'poge' and the cube was thrown up the stepped sides of the pit and passed to the surface by men standing on a series of ledges or 'eaves' stepping up the side of the pit at vertical intervals of about 5ft. Examples of the tools used are shown in photo **6.9**. Water from cans, often seen in old photographs, was used to lubricate the tools, and also the clay balls as they were slid across the clay benches. At the surface the balls were loaded onto a cart. Alex Scott, writing in 1929, noted that the process of cutting the 'balls' was as described above, but the clay was taken to the

surface in tubs along inclined tramways. Once the floor of the pit was cleared, the process from thirting to poging would be repeated. Photo **6.8** shows an example of a pit being worked in this way.

In the early 1930s hand-held pneumatic spades were introduced to replace the thirting iron, but otherwise the methods used were much the same, with the clay being still cut into cubes. Lumpers continued to be used to undercut the cubes. Pneumatic spades were used first by Newton Abbot Clays, then by Devon & Courtenay and the North Devon Clay Co.

Square pits

The soft sides of open pits were prone to collapse so that timbers began to be used to support their sides – thus giving rise to the timber-lined and braced 'square pits' of the Bovey Basin. The square pits were generally from 18ft to 24ft square, occasionally, but rarely, sunk to depths of 100ft. While 'Stoneware' clays continued to be dug from shallow open pits, the more valuable 'Whiteware' clays that typically occurred at deeper depths were at first (until the development of shaft mines) dug using square pits. The square pits were heavily timbered since they were often sunk through unstable ground, particularly the overburden of gravel and sand ('heading') which commonly overlies the clays. Cyril Brackenbury, writing in 1931, noted that in the larger pits the strong timber braces, called 'stays' were 12in square, but in smaller pits round logs were used. These stays were wedged tightly at the ends against wooden blocks called 'gear-pieces'. These were placed against two 'stay-boards', 12in wide and 5in thick, which were placed end-to-end and extended the full length of the side of the pit.

Sometimes bundles of brushwood, heather, or green foliage (called 'wreathes' or 'vraiths') were pushed against the pit walls behind the stay-boards, to help support weak walls, or to stop running sand flowing in to the pit. The inflow of water into these square pits was unpredictable; some stayed completely dry, while others might have to be given up if too much water came in for the

6.8. *This photo of an open pit in the Bovey Basin in about 1930 shows the working methods using traditional tools. The man on the right is using a* thirting iron *to cut the vertical sides of the ball clay cubes; in the centre the cubes are undercut with a* lumper; *and the man on the left is spiking the cubes with a* poge *and loading them on to a wagon. Examples of these tools are pictured in photo* **6.9**. Photo courtesy of the Ball Clay Heritage Society.

pumps to cope. The pumps in the early days were rather inefficient, consisting of barrels bored out of a solid trunk of elm; they only had a maximum lift of about 15ft, so that in deeper pits a series of pumps would have to be used. The workers cut gutters (called 'paces') down the sides of the square pit to channel water into a sump from which the elm pumps sucked it out.

A space of between 6ft and 12ft was left on all sides between adjacent square pits. When the clay had been extracted from a pit, it was filled with gravel and sand from the 'heading' and allowed to settle. In some cases, square-pit working took the form of a double row of ten square pits with 18-ft sides and a 6-ft space between each pit. The proportion of the area lost using this method was about a

6.9. Hand tools used in the working of ball clays in the Bovey Basin. Similar tools, but with different names, were used in the Petrockstowe Basin (page 129). In order clockwise from top left: Thirting iron: a kind of spade used to cut the vertical sides of a 'ball' of clay; 5ft long and weighing 10 lbs. Lumper: used to undercut a cube of ball clay; 3ft 7in long and weighing 16lb. Tubil: used to cut clay underground and to trim the sides of workings; 3ft long and weighing 7lb. Poge: used to move the balls of clay; 3ft 4in long and weighing 1lb. Images courtesy of the Ball Clay Heritage Society.

third of the whole area. Where 12ft square banks of clay had been left between the original square pits, these were then extracted in the same way as square pit working, except that the pits were smaller, less than 12ft x 24ft. This process was called 'working the foursides'. In cases where the space between the pits was only 6ft, the workers did not return to work out the clay banks between the square pits.

The clays in the square pits were worked in the same way as small open pits, using the same tools and methods (see page 118 above for a description). The sides of the square pit were of course vertical, and there were no steps or eaves to provide intermediate levels up which the clay balls could be thrown. Thus the clay balls and waste material were loaded into wooden buckets and hauled up from the pit using a crane and wire rope. The simple pivoting wooden jib crane, pictured in photo 6.10, was called a 'crab'. Such cranes were also used to haul material up from shafts.

Each square pit produced a few hundred tons of the several types of clay through it was sunk, and most of the pits were worked for only a few months until the inflow of water became too much for the hand pumps to cope.

Shafts

In the Bovey Basin the square pit system evolved into true underground (shaft) mining by stages; at first the claycutters began to work out from the base of the square pit, giving it a bell-like shape rather like the bell pits by which coal was dug in some coalfields. This method was soon improved on by driving out short timbered levels from the base of the pit. Extraction then became focussed on digging out a particular seam, unlike square pits which exploited a variety of clays as they were deepened through a variable clay sequence. This meant a change from a wide square pit to the sinking of a vertical shaft just wide enough to fit in the necessary bucket, ladders and pump lines.

The discovery of an especially good quality seam of white-firing clay at Kingsteignton stimulated the development of shaft mining there. By the 1870s, in both the Bovey Basin and Petrockstowe Basin, shaft mining was enabling 'potters' clays as deep as 200ft to be worked, especially as the use of Cornish pumps reduced the problems of water influx. The shallower stoneware clays were still worked by open pits and square pits.

A heavily timbered shaft, rectangular in shape, was dug to the depth where the required clay bed or beds occurred; such shafts were

6.10. This photo of a typical timbered square pit shows the simple crane or 'crab' for hauling the clay out of the working, and a pile of clay balls. The crane was held in place by two long legs called 'tie backs'. Devon and Courtenay, Vale Road, Decoy. Photo courtesy of the Ball Clay Heritage Society.

typically 50 to 150 ft deep, the greatest depth reached being about 200ft. The sides of the shaft were supported by horizontal frames of larch separated by vertical timber 'studdles' between which boards and masses of grass or sedge ('vraiths' or 'wreathes') were stuffed to prevent sand filtering into the shaft. The sinking of the shaft was usually achieved without problems, except where 'running sand' was encountered, which could distort and even destroy the timbering of the shaft. The shaft was divided by timbers into two parts, one of which was used for hauling the clay to the surface, and the other for the miners to reach the workings along a series of ladders placed at intervals, and for the pump lines. The surface appearance of a typical shaft mine is recorded in illustration 6.11, reproduced from the painting by Frederick Cox.

On reaching the desired clay bed, a sump was made at the base of the shaft, and two side tunnels ('drives') were dug in opposite directions from the shaft base for up to 100ft along the level 'strike' of the seam. The drives were normally about 6ft high and 4 to 8ft wide and were closely supported by 6 to 8in-diameter larch poles. The amount of timbering depended on the type of clay: a plastic clay needed timber supports at intervals of 2ft or more, but a 'short' (less plastic) clay needed closer timbering, at intervals of 1ft, being more liable to collapse. Working of the seam then took place by cutting a series of adjacent drives (in the numbered order shown on **6.12**) following the inclined clay seam upwards ('on the rise'), in a fan-shaped pattern, until they met. At the start of each new drive, the timber was removed from the previous drive and re-used. If a thick

6.11. A typical shaft mine in the Bovey Basin. Reproduced from a painting by Frederick Cox, 1995, with kind permission of the artist.

Right: *6.12. This drawing shows the 'fan' method of ball clay mining in the Bovey Basin.* From Brackenbury, 1931, figure 46.

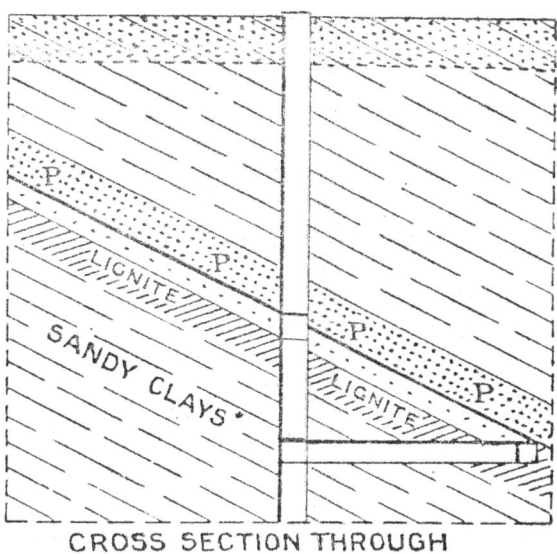

CROSS SECTION THROUGH SHAFT

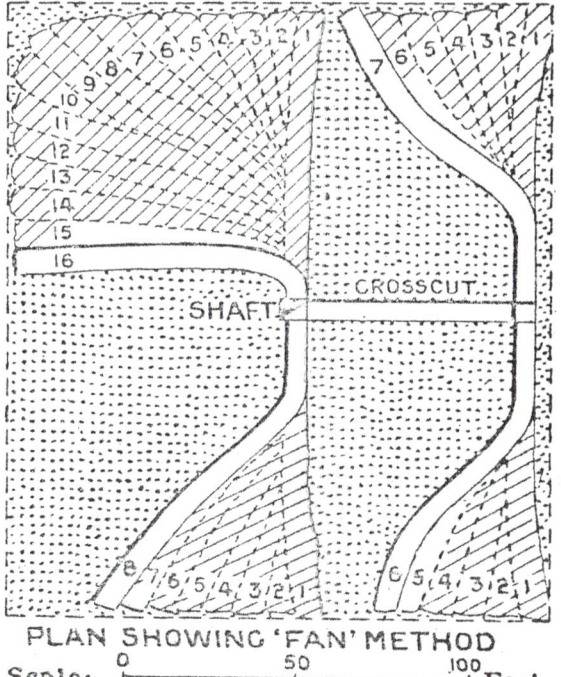

PLAN SHOWING 'FAN' METHOD

Scale: 0   50   100 Feet

6.13. Working at the face in a shaft mine in the Bovey Basin in 1951 (Bill Tribble's Shaft, Southacre, Preston Manor, Kingsteignton). The clay is being cut by Geoff Gibbs using a pneumatic spade, while Les Bickham is loading the clay onto a barrow. Note that the larch roof timbers are slowly cracking under the pressure from above. Photo courtesy of the Ball Clay Heritage Society.

clay bed (15 to 20ft) was being dug, sometimes three or four drives were dug consecutively in the same direction, with more of the clay being dug from the roof in each drive.

If a clay seam was worked from a single shaft without a cross-cut, the area worked was not normally greater than about 180ft by 90ft. However, if the seam was also worked from a lower level by a cross-cut, as shown in **6.12**, the total area worked was increased to about 180ft by 160ft.

The clay was loaded into barrows at the face (**6.13**), pushed by hand to the base of the shaft, and hauled to the surface in buckets. In earlier times the buckets were pulled to the surface using a hand windlass, but later horses were used. In the late 1920s, the ropes were operated by mechanical or electrical means. There were no rails along the floors of the levels, but dry clay from the surface was used to give a good surface underfoot.

One particular and notable feature of ball clay mining, particularly at shallower depths, was the effect of the plasticity of the clays, which behaved almost as if they were alive. The clay would, over even quite short periods of time, flow into the workings, the

floors would heave, and the clay squeeze from between the timber supports. When supports were removed, a tunnel could close up completely in only a few days, such was the plasticity of some clays. Photo **6.14** shows a mine tunnel in which the wooden timber supports have been cracked and distorted by the squeezing action of the surrounding plastic clays. Most shafts had a life of no more than two or three years.

Where the workings were subject to inflow of water, Cornish pumps were used to dewater them, but many shafts were dry enough for this not to be needed. Ventilation to the levels was provided by a fan which worked intermittently, but occasionally in shallow shafts the only means of ventilation was the return down the shafts of part of the water pumped up. Even in the mid-1920s the miners still used candles for illumination, but in a few cases electric hand lamps were used.

A typical shaft of the type described above was operated by a gang of six men. Four were involved in the actual digging or 'getting' of the clay, while two men worked on the surface. Output per shaft was about 18 tons a day. After all the clay within about a 60 to 90 ft radius of the base of the shaft had been dug out, it was abandoned. Only four miners worked underground at a time. At each of the two 'headings' a miner cut the clay with a tubil (pictured in **6.9**). A second miner in each heading used an elm wheelbarrow (with a capacity of about 3cwt - 336lb) to take the clay to the base of the shaft. A 'top ganger' who was in charge of the gang operated the crab crane at the top of the shaft and the clay was loaded from the crane bucket into wagons. A 'top trammer' was in charge of hauling the wagons up elevated timber ramps or 'high backs' from where the clay was dumped into separate heaps. The screeching of the ubiquitous wire ropes in their pulleys was the characteristic noise of the ball clay works.

Even as late as the mid-1920s, little seemed to have changed in mining methods, although some electrical and mechanical improvements were beginning to be introduced. Clay continued to be cut by hand using the tubil for many years, despite the difficulty of using it in confined spaces. Mechanical tools in the form of small

Opposite: *6.14. A ball clay tunnel showing the effect of the movement of plastic clay on the timber supports which have been twisted and broken by the slow movement of the clay.* Photo courtesy of the Ball Clay Heritage Society.

pneumatic spades were introduced in 1932 to the Bovey Basin for underground clay-cutting, but only came into universal use after the Second World War. Timber continued to be used for supporting headings until 1951 when steel arches were introduced by the Devon & Courtenay Clay Co in a mine at West Golds, near Newton Abbot.

Another method of working which differed from the 'fan' method described above was by main drives, secondary drives and offshoots (**6.15**). It was described by the mining engineer Cyril Brackenbury in 1931 and in fact developed by him in order to increase the area of ground which could be worked from each shaft. It was suitable for working clay beds of medium thickness which were not dipping too steeply. It was also especially suited to areas where the ground conditions were difficult and therefore costly for shaft-sinking; such an area was that at the southern end of the Bovey Basin in the tidal part of the River Teign, where the clay beds were overlain by water-

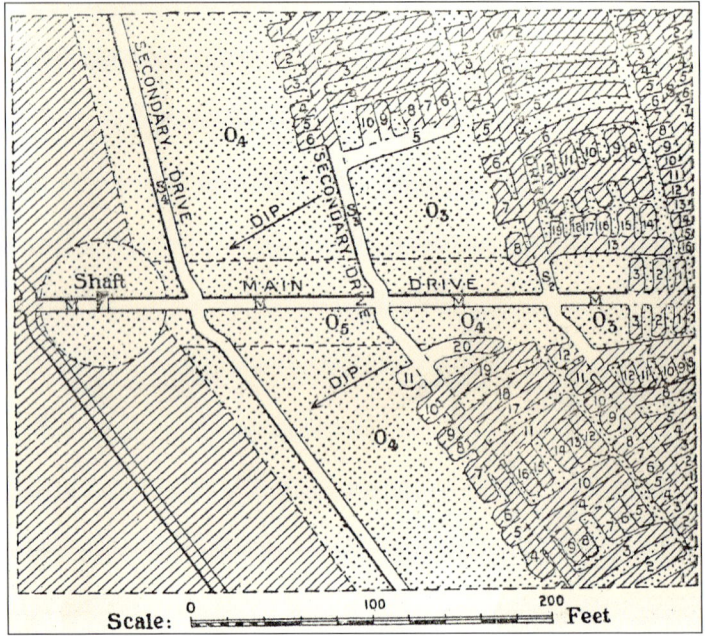

6.15. *This drawing shows a method of mining ball clay by main drives, secondary drives, and offshoots.* From Brackenbury, 1931, figure 47.

logged sands. Brackenbury developed a special method of sinking shafts through this type of ground.

Firstly a long main drive (M in illustration **6.15.**) was made from the base of the shaft, on the 'rise' of the clay bed that is, following its dip upwards. From this, level secondary drives (S$_1$, S$_2$, S$_3$, S$_4$) were made along the strike of the clay bed, and lastly 'off-shoots' were made at right angles from the secondary drives. The off-shoots were longer on the rise of the bed and shorter down the dip of the bed. The numbers on the off-shoots show the order in which they were worked out. On the figure (**6.15**) the areas worked out are shown by diagonal lines, and the areas shown dotted and lettered O$_3$, O$_4$, O$_5$ are the areas of clay bed still to be worked. A solid pillar of clay was left around the base of the shaft to provide support for it. Work began at the farthest end of the main drive so that the clay around the shaft was the last to be dug.

Adits

Although a few small primitive adits (photo **6.16**) had been driven in from the edges of some open pits in the 1920s and 1930s in the Bovey Basin by Devon & Courtenay, it was during the 1960s that clay-mining from modern inclined adits became the only method of underground working, completely replacing the vertical shaft method. In this method, an inclined tunnel was driven at a shallow angle of about 1 in 2, or 1 in 3, from the surface east of the outcrop of a clay seam, until it intersected the required clay bed underground; it then followed the clay bed at dips varying from 1 in 4, to 1 in 8, or 1 in 10. These adits were first tried experimentally in 1957-58 and came into widespread use in the 1960s. They enabled clay to be got from vertical depths of up to 400ft below the surface.

The first experimental modern adit was driven by Watts Blake Bearne in 1957-58. It was driven for 220ft from the floor of an open pit at a gradient of 1 in 2 until the required clay seam, about 6ft thick, was reached. From that point, the adit followed the dip of the clay bed which was 1 in 4, then easing to 1 in 8. At a length of 1000ft the depth below surface was 250ft. The tunnels were 6ft high and 8-9ft wide, supported by timbers spaced at 15in apart. An interesting point is that, as deeper levels were reached, especially below 150ft, the ground became more stable and the clays less plastic. L T C Rolt has amusingly compared clay mining at these depths to excavating dark chocolate, which the clay resembled in

6.16. *The entrance to a primitive adit mine at Little Bradley Pit, Chudleigh Knighton, about 1948-49. The figure on the right is Fred 'Major' Harvey.*
Photo courtesy of the Ball Clay Heritage Society.

colour, friability and in its slightly greasy surface texture. But, on exposure to air, the 'chocolate' faded to a putty colour. The firm 'chocolate' clay dug from the deeper mines was compared to the shallower mines where the more plastic clays were said to be like 'blancmange'!

There were several reasons why adit mining developed. Deep boreholes had proved the existence of good quality clays at depths too deep for shaft mining. Also, the Mines & Quarries Act (1954) made it illegal to use ladders in shaft mines over 150ft deep, and it was not only uneconomic to install mechanical man-riding equipment in ball clay shaft mines because of their short life and small scale, but also impractical because of the distortion that the shafts suffered owing to the plastic nature of ball clay. These factors prompted the development of inclined adit mines.

During the adit mining period in the Bovey Basin, Group 1 clays (page 112) were extracted from the West Gold Marshes No. 10 Adit (Watts Blake Bearne & Co Ltd) and the Broadway Nos 7/8 Adits (ECC Ball Clays). Group 2 clays (page 112) were extensively mined from Rixeypark Mine (No 11 Adit – pictured in photo 6.17), Preston Manor No. 5 Adit, and from Preston Manor No. 4 Adit (a plan is shown in 6.18), all then operated by Watts Blake Bearne & Co Ltd. Group 3 clays (page 112) were dug during the adit period

from Rixeypark Mine (No. 11 Adit) and Preston Manor No. 5 Adit (both then operated by Watts Blake Bearne & Co Ltd).

Tony Vincent, formerly Chief Geologist of Watts Blake Bearne & Co Ltd, notes that (in the case of that company) the modern underground adit mines were developed to supply essential components (for example 'Best Light Blue' and 'Dark Brakes') to be used for the blends which could not be dug in sufficient quantities from the open pits. The clays from the adits were in fact produced at a loss in terms of cost per ton of clay mined. The last adit to close was ECC Ball Clay's No 8 Adit at Broadway which outlived the Broadway open pit by a decade or so.

An adit mine (No 11) in use in the late 1960s is pictured in photo 6.17. A 1964 article in the *Quarry Managers' Journal* gave interesting details about the operation of a similar inclined adit mine (the No 4 Adit of Watts Blake Bearne & Co Ltd), which was located in the Southacre area of the Bovey Basin in the mid-1960s. A plan of the adit is given in 6.18. It had been driven to a depth of nearly 400ft

6.17. A photo of an adit mine in the Bovey Basin in the late 1960s. It shows No 11 Adit (Rixeypark Mine) {SX 8507 7638}, operated by Watts Blake Bearne & Co Ltd. Photo courtesy of the Ball Clay Heritage Society.

126

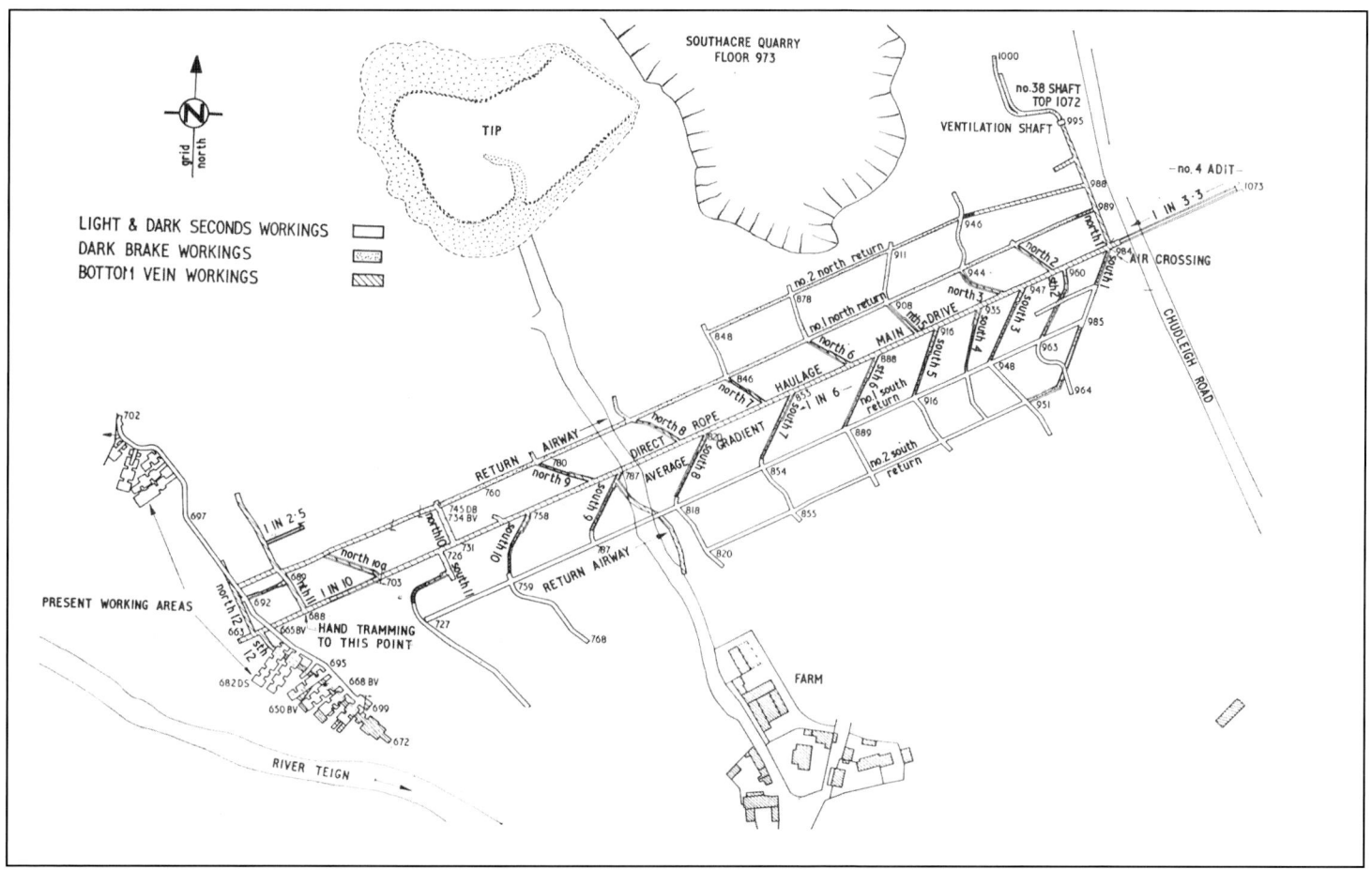

6.18. A plan of No 4 adit mine, located just south of the Southacre Pit in the Bovey Basin. Reproduced from *The Quarry Managers' Journal*, August 1964, figure 7, with permission.

in a seam called the Bottom Vein (this is part of the 'Group 2' Clays, described on page 112). Above the Bottom Vein are two other clay seams, the 'Dark Brake' and the 'Top and Bottom Light and Dark Seconds' which were worked as one seam (illustration 6.19 shows a section through the seams worked in this mine). From the adit, which was the main haulage way, sub-level drives were turned off to the north and south to intersect the two seams above. The drives were developed at an angle of about 45° to the adit and firstly went

through about 8ft of sandy clay between the Bottom Vein and the Dark Brake (see 6.19). When the good clay was reached, the levels to the north were turned and driven in the seam to open up a working area, The levels south of the adit were driven through the Dark Brake in to the Light and Dark Seconds where again development was continued on the seam horizon. On the mine plan, 6.18, it can be seen that in the upper section of the mine, the clays were blocked out by driving sub-levels to the south and north and connecting these with return raises running parallel to the main adit.

The miners worked in pairs, using compressed-air spades to dig out the clay from the working face. Later in the 1960s, clay-cutting machines such as that pictured in photo **6.20** came into use. The clay was loaded into 12cwt mine-cars which were pushed by hand to the main adit, where the cars were attached to a haulage rope and pulled to the surface. They were then hoisted and tipped into the clay storage shed. Each team produced on average about 7.5 tons per man-shift.

Support for the workings described in 1964 was provided by timbers spaced at 18in intervals. A square timber was laid on the floor, then two vertical posts against the side of the drive. A fourth timber was then placed on top of the vertical timbers to support the roof of the tunnel. However, steel D-arch supports began to be used in the late 1960s, as shown in photo **6.20**.

The Petrockstowe Basin

Mining has taken place in two main areas of the Petrockstowe Basin: to the north in the Marland Moor area west of Merton; and in the southern area west of Meeth (see the map, **6.4**). The well known Marland Brick and Tile works was located in the northern part of the basin at Clay Moor.

Marland

The exact date when the Marland clays were first worked is uncertain, but the Greening family of Bideford, who owned land in Petersmarland and Petrockstowe parishes in the second half of the 17th century, may have shipped clay from there. Clay from Clay Moor was being dug in the 1850s and 1860s. In July 1875 a visitor to Clay Moor reported a shaft sunk to nearly 80ft depth, probably for

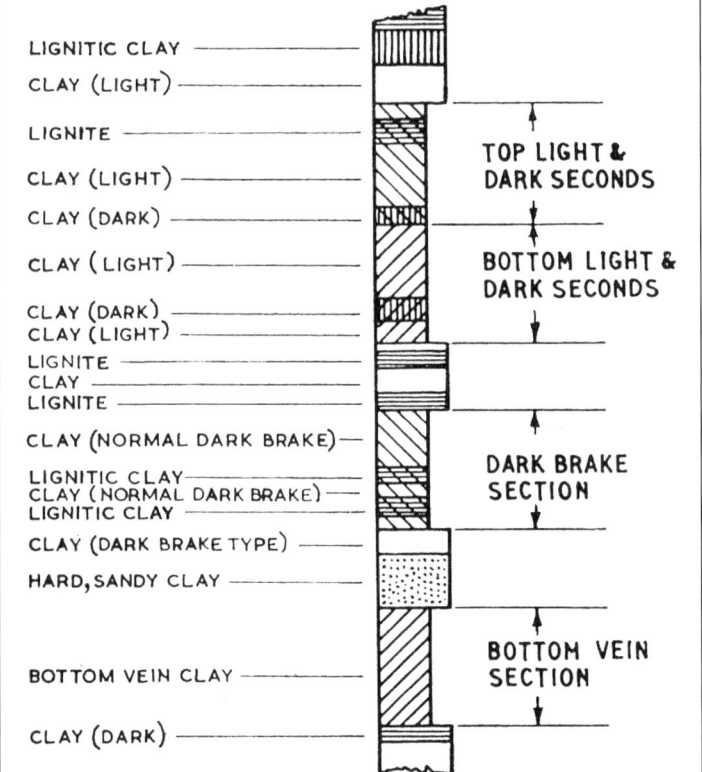

Top right: *6.19. A vertical section through the seams worked in No 4 adit mine located just south of the Southacre Pit in the Bovey Basin.* Reproduced from The *Quarry Managers' Journal*, August 1964, figure 8, with permission.

Right: *6.20. Men working at the face in a Watts Blake Bearne adit mine, Bovey Basin, about 1967. The clay is cut at the face by a boom with rotating knives and taken by conveyor to the wagon. Compare the working conditions with those in a 1951 shaft mine (6.13). The miner is Ivor Basset.* Photo courtesy of the Ball Clay Heritage Society.

exploratory purposes; it had been infilled by November 1877. Clay was being dug from several open pits and by 1878 the kilns and buildings of the Marland Brick & Tile Works had been constructed by William Wren. The North Devon Clay Company worked open pits in the Marland area for supplying the brickworks and other markets, and in the five years to 1890 the annual tonnage of clay dug doubled to over 10,000 tons. By 1892, some open pits were up to 30ft deep and at this depth water became a problem.

The open pits at Marland were worked by the usual open cast methods, as described above (page 118) for the Bovey Basin. However, the names of the special tools are reported by Michael Messenger to have differed from those used in south Devon: the Thirting Iron, Lumper and Poge of the Bovey Basin (see page 118) were in north Devon called, respectively, Cutter, Digger, and Prodder. In the 1920s 40 or 50 men were employed, but this number fell as underground mining took over.

About 1892, the first commercial shafts were sunk at Marland, and by 1897 shaft mining had mostly superseded extraction from open pits. The shafts were sunk at angles of between 55° and 75°, allowing wagons to run underground and be winched to the surface along rails laid in the shafts (6.21 and 6.22). The inclined-shaft method probably continued without much innovation, under the management of the same company, until mining ceased in 1969. A 1948 report on the mines recorded that the shafts were between 20 and 100ft deep, according to the clay bed worked, and the total length of each shaft was about 300ft. The gradients of the shafts varied from 'moderate' to 1 in 3. From the farthest extent of the shaft, the clay was worked to a distance of up to 30ft on each side. Six men formed a team: four underground and two on the surface. Each team of miners was allocated a daily amount of clay to dig, which was known as a 'hag', equivalent to about 38 bucketfuls.

Another account by Ted Freshney of inclined-shaft mining from the late 1960s gives additional information, as follows: when the required clay seam (typically inclined at between 5° and 30°) was reached, a tunnel was driven down the dip of the clay seam as far as possible. From the end of the tunnel, levels were pushed out on either side, at first from the end of the tunnel, then successively towards the base of the shaft. The clay was loaded into tubs which were lowered down the shaft and tunnel using gravity, and hauled up to the surface by electric winding engines. The levels usually extended to about 20ft either side of the main tunnel; when the clay

Top: *6.21. This photo shows shaft mines in the Marland area of the Petrockstowe Basin, about 1920. The figure on the left wearing a cloth cap is John Henry 'Jack' Holwill, pit foreman between the wars, while the man wearing a bowler hat is possibly his father, Henry Holwill.* Photo: North Devon Museum Trust Collection.

Above: *6.22. Shaft mines in the Marland area of the Petrockstowe Basin, about 1930. The locomotive is either* Jersey II, *or* Merton. *The line is a three-feet narrow gauge.* Photo: North Devon Museum Trust Collection.

was worked out, the shaft was sometimes deepened and a further sequence of levels driven in a lower clay seam. In this way up to three

6.24. Digging clay in a Petrockstowe Basin mine (possibly in the Marland area) with pneumatic spades. The date is uncertain. Photo: North Devon Museum Trust Collection.

Left: *6.23. A tunnel in a north Devon ball clay mine, lined with bricks from the Marland brickworks; the bottom part of the face looks dark and lignitic, which suggests a mine in the '116' seam or one of the '120' seams. The location and date are uncertain, but since pneumatic spades did not come into use until the 1930s in north Devon, the photo was probably taken after that date.* Photo courtesy of the Ball Clay Heritage Society.

separate clay seams were worked successively from the same shaft. When the shaft was abandoned, another shaft was sunk about 40ft along the strike of the clay seam, and the whole process was repeated; occasionally, the new workings broke into abandoned old levels from a neighbouring mine. By this means, seams of clay were extracted for up to 300ft from the line of shafts.

In some mines on the southwestern flank of the northeastern shelf, the seams steepened to vertical in the vicinity of the internal fault that is present in the Petrockstowe Basin (see map **6.4**), and became unworkable. One mine foreman reported that they went on in one mine down the vertical seam and by the time they gave up it

had rolled over to dip steeply northeastwards.

At some stage, tunnels in some of the north Devon mines were lined with bricks from the nearby Marland Brick & Tile works, as shown in photo **6.23**. It is not known how widespread this type of support was, nor how successful. It certainly must have been very labour intensive to set the bricks, and large numbers of bricks were needed; possibly they were mainly wasters from brick production, of too poor a quality for sale. Certainly by 1965 the supports in the North Devon Clay Co mines were all of timber.

In the 1930s, the traditional tools were replaced by pneumatic spades, but these were heavy (weighing nearly 55lb) and had to be lifted to head height for each cut of the clay in the cramped heading (see photos **6.23** and **6.24**).

Electricity was introduced to the works and mines in 1936. The mines were ventilated by air pumped through 6in-diameter galvanised tubes. Water was not usually a problem in the north Devon mines, and until 1937 was baled into a tank lowered in the mine bucket. After that date, compressed air pumps were used, at first to fill the tank mentioned, then to pump the water directly to the surface.

In 1967, 94 men produced 35,000 tons of ball clay from nine mines and one small pit, whereas in 1980 35 men produced 100,000 tons of clay from two large pits; 75% of the production was exported. In 1979 production was about 120,000 tonnes per year. By 2007 only 20 men were employed. In 2006, production was over 500,000 tonnes; 40% of output was taken to south Devon for blending with other clays, 40% was exported and 20% sold within the UK.

By the 1970s the clays were worked in open pits in the northeast of the basin by Watts Blake Bearne and Co Ltd, who had two pits at Courtmoor and Westbeare. Today these pits, owned by Sibelco, the successor to Watts Blake Bearne, have been amalgamated into one large open pit. In 2010 there was only one company (Sibelco) working in the Petrockstowe Basin; the other main ball clay company (Imerys Minerals) ceased production in the Petrockstowe Basin in 2004.

Meeth

The ball clays near Meeth were even more remote than those at Marland and consequently not economically viable until transport links improved. With the approaching construction of the North Devon & Cornwall Junction Light Railway (NDCJLR) the ball clay resources of the area became more attractive economically, and the clay rights to Woolladon and Stockleigh Barton were obtained by Eustace Holwill and in 1920 transferred to the Meeth (North Devon) Clay Company Ltd. Clay pits were opened at Woolladon and Stockleigh, with production starting in 1921. The open pits were worked by the usual open cast methods, as described above for Marland and the Bovey Basin. With rising production, the state of the roads deteriorated with increasing ball clay traffic, but with the completion of the NDCJLR, which passed almost a mile from Woolladon, the situation improved, and the pits were connected to

the standard gauge line by a two-feet gauge tramway, construction of which began early in 1924 and was probably complete by March 1925.

In 1926, the Corn and Brain families from the Potteries became controlling shareholders, and their experience in coal mine management led to a change in the direction of the company. It was decided in September 1930 to concentrate on underground mining by the 'footrail' method, illustrated in **6.25**. A shallow incline was driven to the desired clay bed from the surface and at its end tunnels were driven off on each side at right angles. At the end of these tunnels further tunnels were driven at right angles, that is, parallel to the main incline from the surface. At their ends, the ball clay was dug out for 20ft on either side. Once the clay had been removed the process was repeated from another level 60ft nearer to the main incline. Clay was loaded onto wagons on tramways running down into the levels. The wagons were hauled up the main incline by cable, and along the other tunnels were either man hauled or pulled along by winches.

Compressed air and electricity were at first provided by steam-driven plant. but in 1934 the works were connected to the main electricity supply. Work stopped at Woolladon pit when underground mining started. After the Second World War, new mines were opened and Woolladon pit was re-opened.

In the late 1960s underground workings in the southern (Meeth) part of the Petrockstowe Basin were quite different from the northern inclined-shaft workings. The adits were D-shaped, driven at inclinations of about 1 in 3 until the required seam was reached, and then the adit was driven within the seam for up to 660ft. Perhaps the most extensive of these, the No. 1 mine, was a curious affair. The tunnels followed the clay seams which were in places only 10-20ft below the surface, and in places where the seam came to outcrop, the tunnels also came to the surface. The plastic nature of the clays was very evident in this and other mines, and the squeezing was so intense in places that one side of the tunnel collapsed towards the other side and the miners had to crawl through the narrow gap between the walls. From the adit, levels were driven on either side for up to 250ft, and small stall workings driven off the levels. The clay was hauled out in tubs which ran down the main adit under gravity and were pushed by hand along the side levels. Winding gear was installed at some crossroads on the main drives to haul the full tubs from the levels.

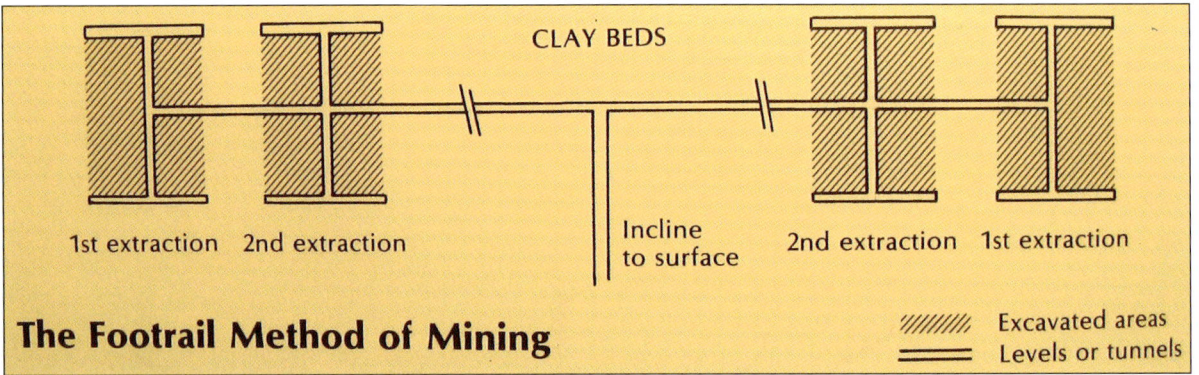

CLAY BEDS

1st extraction   2nd extraction

Incline
to surface

2nd extraction   1st extraction

The Footrail Method of Mining

///////// Excavated areas
======== Levels or tunnels

6.25. The 'footrail' method of mining at Meeth (page 131). From Messenger 2007, page 96, courtesy of the author.

With the new ownership under English China Clays Limited from 1967, there was again a change in policy, as there had been with the change in ownership in 1926. In this case, the opposite happened and the mines were phased out in favour of open pit working. By the mid-1970s ball clays were worked in open pits in the south of the basin by ECC Ball Clays Ltd who had three pits, at Meeth, Woolladon and Stockleigh Moor (the latter opened in 1976). During the 1970s, production of ball clay rose from about 40,000 to 60,000 tons a year. In 1999, following the takeover of ECC by the French company Imerys, production of ball clay continued at levels of around 60,000 tons a year. However, production ceased and the works closed in December 2004 because better quality clays could be dug from the Bovey Basin deposits owned by the company.

HOW BALL CLAYS ARE USED

The most common use for ball clays is in the ceramic industry, and about 80% of the production is used for the manufacture of sanitaryware, floor and wall tiles, and tableware. The economic importance of ball clays is demonstrated by the fact that the value of ball clay production in 2004 was about £46 million. The particular properties which make them so valuable are their plasticity (which means they are easy to mould); their unfired strength; and the fact that they fire in a kiln to a white or near-white colour. Other ceramic uses include refractories, kiln furniture, electrical porcelain, enamels and glazes, and building bricks. Since about the 1950s ball clays have also been used for a variety of non-ceramic purposes, such as

reinforcing fillers in rubber and linoleum, extenders in animal foodstuffs, and for coating fertiliser pellets. Ball clays are not normally used alone, but are generally mixed with other constituents such as china clay (kaolin), silica sand, and a flux. The amount of added material depends on the product.

Ball clay was for many years sold in 'lump' or 'ball' form just as it came from the pit or mine, and stored in long low buildings called 'clay cellars'. In Newton Road, Kingsteignton, by the Hackney canal, are clay cellars built by Lord Clifford in about 1843. Other clays, such as the better-quality white-firing clays, were in these early days stored outside in open 'bedplaces', for it was then thought that weathering would improve the working qualities of the clay. In 1938 a company in the Bovey Basin received a request from America to provide clay in shredded form, that is, cut into pieces about an inch long and stored under cover to reduce moisture and hence transport costs. Apparently an adapted turnip cutter was used to shred the clay for this first order! This innovation caught on, and most ball clay is today produced today in shredded form. Other clays are dried and powdered. In 2009, about 76% of ball clays were sold in shredded form, about 6% were sold in powdered form, and about 18% were refined.

Another process involved calcining the clay by heating in rotary kilns, the resulting product being less reactive and used mainly in the manufacture of refractories. The calcining of clays from the Chudleigh Knighton area for high temperature refractories came to a sudden end in 1979 when clays from the Clay Lane pit (which had been opened up as a substitute for the Chudleigh Knighton pit) became too high in carbon content. Eventually, after it had been 'mothballed' for a number of years, the calciner was demolished.

Blending of clays started in the early 1950s after the introduction of shredders which made blending possible, and continues to this day. Refining became important in the 1970s, particularly for blends of clay used to make sanitaryware. The world's first refined clay ('Sanblend 75'), specifically aimed at the needs of the vitreous china sanitaryware industry, was produced by Watts Blake Bearne & Co Ltd in 1975. Other refined clays are now produced, the latest being 'Sanblend 90'. The blends involve mixing several different types of clay, in some cases from both the Bovey and Petrockstowe basins, to achieve the desired properties and this process require sophisticated selection and quality control techniques.

What of the future of the ball clay industry? There are still large reserves, especially in the Bovey Basin, and Devon will continue to be the major UK source of ball clay in the foreseeable future. In 2011 it was estimated that in the Bovey Basin there were 45 million tonnes of permitted reserves representing 136 years production, and about 19 million tonnes of unpermitted resources in the east of the basin (there are also potential resources in the central and western parts of the basin). In the Petrockstowe Basin there were about 6 million tonnes of permitted reserves representing 23 years production, and 35 million tonnes of permitted but not fully explored resources, representing 138 years production. The British Ceramics Confederation predicts that UK whiteware manufacturing will decline slowly over the coming years as companies continue to invest in overseas production. For tableware it is not uncommon for manufacturers to carry out blank production overseas and then carry out the decorative finish and firing in the UK. Export sales of the ball clays used for sanitaryware, which have a worldwide reputation, have increased over the years, but the more siliceous clays used in the tile industry have seen increasing competition from other sources, such as those in the Ukraine.

TRANSPORTING BALL CLAYS

The Bovey Basin
In the earliest, pre-railway days, ball clay was moved from the pits and mines to the ports by packhorses and carts. For the Bovey Basin the nearest port was Teignmouth and in the case of the Petrockstowe

Basin the main port was Bideford. Dean Milles, writing between 1747 and 1762, recorded that: ' ...they load [the clay] on horse-back and carry it two miles to a place called Hackney in Kingsteignton parish where it is shipped on board vessels for Liverpool from which place it is carried to Staffordshire. The owner of the land sells it at ye waterside for about 7/- per tun'.

Canals were constructed to reduce transport costs. The Stover Canal was built in 1790-92 between Teigngrace and the Teign Estuary. The canal was born from the need to transport ball clay, but was also used for a time (at various dates between 1820 and 1858) to transport granite from Haytor on Dartmoor (see page 145 for more about the Haytor Granite Tramway). The Hackney Canal between Kingsteignton and Hackney Quay opened in 1843. Various clay cellars were built beside both canals to store clay before loading on to ships for onward transport. M Dunsford noted in 1800: '... the conveyance of clay ...to Liverpool, for the fine earthen manufacture of Staffordshire; it is dug in the parish of Kingsteignton only, formed into little irregular squares somewhat like bricks, weighing thirty-six pounds each. About seven thousand tons are annually sent from these pits, in barges on Mr Templer's Canal, to the River Teign, and down that river to the port, at 1s. 6d. per ton, toll and boatage, and there dispersed to the several manufacturers of Staffordshire...'

In about 1858, the Teignbridge clay cellars were sold by the Duke of Somerset to the Moretonhampstead and South Devon Railway, together with the canal and the strip of land for the Moretonhampstead branch railway line. Watts Blake Bearne & Co Ltd leased the canal and the clay cellars from the South Devon Railway/Great Western Railway from 1867 to 1942. The clays for transhipment at Teignmouth were taken by barge via the Stover Canal, Whitelake, and the Teign Estuary, the rail links not being used for that purpose. Clays from the Decoy Basin (Devon & Courtenay) were loaded onto barges at Newton Abbot Quay. Thus for a century of more the canals and the Teign Estuary were the main transport routes for clay transport.

To transfer the clay from the canal barges to the coasting vessels, a long pole-like tool was used to stab the 'torbs' or cuboidal lumps of clay which were then thrown back over the workman's shoulder up into a chute leading to the hold of the coasting vessel. The work was normally carried out by eight men called 'lumpers', and M C Ewans notes that 'with 13 tons to shift every day they must have fully earned their pay, which was only four to five shillings a day'.

The coming of the railways was naturally of great advantage to an industry that needed to move a bulky product cheaply and efficiently to its markets. The South Devon Railway was opened to Teignmouth in May 1846 and to Newton Abbot by the end of 1846, but the railway company concentrated on passenger traffic and the ball clay industry was not greatly affected. It was not until the construction of the branch line to Moretonhampstead (opened in July 1866), which ran through the heart of the Bovey Basin, that the railway began to be widely used to carry ball clay. Clay could be loaded at Heathfield station from 1874, at Teignbridge from 1890 and at East Golds from 1938.

Today most clay for use within the UK is transported by road, especially to Staffordshire. Bulk transport using road tankers was used for ground amine-coated china clay which was brought to the Preston Manor works of Watts Blake Bearne in the Bovey Basin from china clay pits on southwest Dartmoor. This product was used as a carrier for fertiliser. Later, some of the refined blended clays produced by Watts Blake Bearne for sanitaryware (their 'Sanblend 80') were transported by tankers in slurry form. Clay for export to Europe is taken by road to Teignmouth or Bideford and loaded onto ships at the docks. For export to more distant markets, the clay is loaded into containers which are taken by lorry to the major container ports such as Liverpool, Southampton, Ipswich and Felixstowe.

There is a much fuller account of the transport of ball clays by water and land in Chapters 4 and 5 of L T C Rolt's book *The Potters' Field* on the history of the south Devon ball clay industry.

The Petrockstowe Basin

Because of the relatively remote location of the Petrockstowe Basin ball clays (compared, for example, to the Bovey Basin deposits), transport has always been a vital factor influencing the development of the industry. Although the north Devon clays had a wide market in the 17th and 18th centuries (and possibly even the sixteenth), during the 19th century the lack of good communications to Marland and Meeth hindered development of the ball clay resources there. As we have seen, Lysons reported that clay pits in the Petrockstowe Basin had not been worked for 20 years before 1822, possibly due to competition from the Bovey Basin pits which had better transport links. However, the situation improved in 1827 with the opening of the Rolle Canal which connected Torrington with Annery on the River Torridge south of Bideford.

By the early years of the 20th century, clay sales were increasing steadily. North Devon clay was shipped through Bideford and Fremington, with a significant export trade to the continent and America; clay for America usually went via Liverpool. The British market was served by coastal vessels sailing to the Mersey for transfer on to canal boats which took it to its final destination in the Potteries.

In the late 1870s, the owner of the Marland brickworks, William Wren, realised the importance of transport to the future of the clay industry, and financed the construction of a light railway (the Torrington & Marland Railway) 6¼ miles long, from Torrington to the Marland works. The line, built by John Barraclough Fell, opened in January 1881; by 1886 it had been extended to Bury Moor at the southern limit of the company's land. In order to make the line economical to build and run, it was constructed to a three-feet gauge. Money was also saved by gaining the consent of all the landowners along the route, thus avoiding the cost of obtaining an Act of Parliament. Another distinctive feature, designed to save money and to speed up construction, was the use of wooden viaducts which saved the expense of constructing embankments and bridges. There were six such structures along the line, the longest and most impressive of which was the 316 yards-long viaduct across the River Torridge just south of Torrington station. Within the ball clay country south of the Marland works there were many short branches from the line which served individual mines and pits.

Various proposals for a standard gauge railway linking Torrington and Okehampton and passing through the Petrockstowe Basin were made between 1831 and 1895 but none came to fruition. In 1909 a proposal was made by Holman Fred Stephens to build a standard gauge light railway, the North Devon & Cornwall Junction Light Railway (NDCJLR) from Torrington to Halwill Junction, a distance of 20½ miles. With the opening of the new line in 1925, the Torrington and Marland Railway had, in the words of Michael Messenger, 'ceased to be anything more than an internal works railway.'

The First World War intervened, and with further delays, construction of the NDCJLR did not begin until begin until 1922. The line opened on 27 July 1925. Not surprisingly in retrospect, it had opened too late and never achieved great passenger and goods figures; competition from road transport had already begun.

Although the Marland brickworks were re-opened in 1925, clay production from Marland and Meeth did not reach the tonnages expected for many years. On 7 September 1964 the line closed to goods traffic except for clay, and on 1 March the inevitable happened and the passenger service was withdrawn. Clay traffic continued, but the closure by British Rail in September 1969 of cranes at Fremington Quay meant that clay could not be loaded on to ships there. Clay continued to be transported by rail, to Bideford, Teignmouth, Fowey, or to Stoke-on-Trent, but most was transported by road. By 1982 the life of the railway was over: the last trains carrying clay left Meeth on 23 August and Marland on 13 September 1982. Of the production of ball clay of between 40 and 60,000 tons from Meeth in the 1970s, about half was exported through Bideford, Teignmouth and Fowey. Until the railway closed in 1982 about a third was sent by rail to Stoke-on-Trent. After that date all transport was by road.

The Torrington & Marland Railway continued serving the clay workings, with the rails being moved to fit in with the positions of new workings. However, by the 1970s, when open pits had replaced the shaft mines, the railway was less well adapted to serving the continuously changing working face of a pit rather than a static pithead, and it closed on 6 November 1971. The full histories of the Torrington & Marland Railway and of the NDCJLR are given in the 2007 book *North Devon Clay* by Michael Messenger.

In 2006, most of the export production from Petrockstowe went via Bideford, transported to the port in lorries, while some went to Plymouth and Teignmouth. A particularly large shipment in November 2006, of 8,600 tons of ball clay from Bideford, was loaded over a three day period onto three ships of the largest draught that could reach Bideford Quay.

Chapter 7
BOVEY COAL:
The Lignite Mines of Bovey Tracey

INTRODUCTION

In this final chapter of the book we stay in the Bovey Basin that we explored in the previous chapter, for it is also within the Bovey Formation deposits there that we find the next valuable commodity that has been won from underground mines, as well as an open pit. This material, called lignite (photo 7.1), is a type of low-grade or brown coal. Devon has virtually the only significant deposits of lignite in the British Isles (there are also sizeable deposits in Northern Ireland). Lignites occur in several areas of the Bovey Basin, but the main place where they have been worked on an extensive scale is around Bovey Tracey, at the northern end of the Bovey Basin (see the map, 6.3). They have been dug not only in a large open pit (the 'Coal Pit' or 'Blue Waters Mine') south of Bovey Tracey, but have also been mined from underground workings. The Bovey lignites have been worked mainly as a source of fuel, but have also been considered as a source of organic chemicals such as montan wax. The nature and the history and methods of working of the lignites form the subject of this chapter.

When we think of coal in Britain we generally associate it with the great coalfields of Carboniferous age (about 320 million years ago) found in South Wales, the Midlands, northern England and elsewhere, which provided the raw materials for the great technological changes of the Industrial Revolution in the 19th century. However there is another kind of coal, widespread in some parts of the world and widely worked in Germany, for example, and

7.1. A piece of lignite in spoil heaps from the old Bovey 'Coal Pit' near Bovey Tracey. It normally occurs in layers or seams alternating with beds of clay.

this is brown coal, or lignite. These deposits are generally much younger than the Carboniferous coals, and many are of Tertiary age (no older than 65 million years). Organic-rich deposits form a continuum which ranges from soft peat at one end, through lignite in the middle, to true hard coal at the other end. The difference is mainly in the moisture and carbon content, and the degree of alteration (or metamorphism) that occurs as an organic deposit matures from peat to anthracite is referred to as the 'rank' of the coal. Peat, for example, has a carbon content of 50-60%; lignite 65-70%; and bituminous coal 75-85%. The highest rank of coal is anthracite which has a carbon content as high as 95%. We have already come across anthracites when we explored the culm mines of north Devon in Chapter 3.

As we have seen, also in Chapter 3, most coal in the coalfields of Carboniferous age formed by the growth and accumulation in place of plant material, and beneath each coal seam there is generally a fossil soil, or 'seatearth', which commonly contains rootlets that extended down from the plants that once grew above. By contrast, the great bulk of the Bovey lignites are the fossilised remains of drifted masses of tree-trunks (mostly of the giant redwood tree, *Sequioa*), washed into a lake as tangled masses derived from forests cloaking the slopes around the basin. However, there are a very few places in the Bovey Basin where beds with rootlets have been recorded, indicating that at least some vegetation grew in place.

Lignite beds are present in many areas of the Bovey Basin, for example, along the eastern side of the main basin in the Southacre Clay and Lignite Member of the Bovey Formation (shown on the map, **6.3**) where they are interbedded with beds of valuable ball clays (described in Chapter 6). A thick lignite bed in the Southacre area was known locally as the 'Big Coal'. These lignite beds from the eastern side of the Bovey Basin have been worked at times. For example, during two months of the coal miners' strike of 1926, 172 tons of lignite were taken from the New Cross Mine of the Hexter and Budge Company, for use in maintaining steam on pumping plant.

At the time of the completion of the 1976 1:50,000-scale Geological Survey map (Sheet 339) it was believed that the lignite sequence at Bovey Tracey was equivalent to the Southacre Clay and Lignite, and that is the interpretation shown on the published map. However, following the results of later intensive drilling of boreholes by the ball clay companies in the western Bovey Basin, it soon

became apparent that the clays and lignites of the Blue Waters Mine area were a younger sequence, named the 'Brimley Member' by Tony Vincent (former Chief Geologist of the ball clay company Watts Blake Bearne) in 1974, and shown on the geological map (**6.3**).

In the central part of the Bovey Basin, at Heathfield, a borehole sunk in the eastern part of the clay pit showed about 260ft of sand and clay on about 200ft of clay and lignite, the lignite in beds up to 34ft thick. The clay and lignite sequence probably belongs to the Brimley Member mentioned above, and is equivalent to the strata at the former Bovey Coal Pit. The Heathfield Shaft, which was sunk in 1924-25 southeast of the old Heathfield railway station as part of an abortive attempt to mine lignite, showed a similar sequence, with lignite between depths of 263 and 283ft. I have come across references to a 'Ligno-Carbon' (or 'Lignite-Carbon') Company (for example Aubrey Strahan in his account of the lignites, describing the Heathfield Pit of Candy and Co, notes that 'The material for the Lignite-Carbon Co. (extinct) was furnished from this pit'). Another source refers to the sale in 1910 of about 13 acres of land which included the premises of the Ligno-Carbon Co, opposite the premises of Candy and Co. I have not been able to establish precisely what the company produced, although the name clearly points to a lignite-related product. This seems surprising, since remarkably little lignite was present in the Heathfield Pit: a section through it, given by Strahan, shows only a single 1ft-thick bed of lignite in a total of about 72ft of strata.

During the 1960s, lignite was extracted as a by-product of ball clay production. Most of it was regarded as waste, but some was hand-sorted and sold for about £4 per ton. The lignite was used for horticultural purposes, as an additive to drilling mud, and for embedding fibreglass swimming pools. In the late 1960s and 70s, lignite from the East Golds Pit near Newton Abbot was taken to Buckfastleigh where it was pulverised and mixed with chemicals. The resulting product, called Acta Bacta, was popular with gardeners as a soil conditioner, but as far as I can ascertain it is no longer available.

THE BOVEY 'COAL'

Despite some early controversy about the nature of the lignite, there is no doubt that it is vegetable in origin, consisting mainly of the remains of trees which have been transported by rivers from outside the Bovey Basin and drifted into their present positions. The regular interlayering of lignite and clay seen in many parts of the basin, including the Blue Waters Mine, suggests a possible seasonal control of deposition.

There are no modern analyses of the lignite, but in 1946 C M Cawley and J G King stated that the lignites have high moisture contents of up to 45%; calculations on dry lignite showed carbon contents varying between 62 and 72%, oxygen contents of 20 to 35%, and hydrogen content of about 5.7%. Bovey lignite straight from the mine had a calorific value of 6,000 British Thermal Units per pound. In comparison, typical good quality coal has a value of 14,000 British Thermal Units per pound. However, William Pengelly, writing in 1863, noted that 'Under careful management, it is estimated that about six tons of [lignite] will produce the heat of one ton of ordinary coal'.

Dr Jeremiah Milles, writing in 1760, recognised differences in the form, colour and texture of different 'coal' beds. He noted that 'The exterior parts, which lie nearest to the clay, have a greater mixture of earth, and are generally of a dark brown, or chocolate colour; some of them appear like a mass of coal and earth mixed; others have a laminous texture, but the laminae run in such oblique, waving, and undulating forms, that they bear a strong resemblance to the roots of trees'. This type of lignite was called '*root coal*' by William Pengelly in 1862.

The 'coals' found in the middle part of the sequence of strata, and in the lowest and thickest bed, were more compact and solid. According to Dr Milles 'these are as black, and almost as heavy as pit coal; they do not so easily divide into laminae, and seem to be more strongly impregnated with bitumen; these are distinguished by the name of *stone coals*, and the fire of them is more strong and lasting than that of other veins'. He continues: 'But the most remarkable and curious vein in these strata is that, which they call the *wood coal*, or *board coal*, from the resemblance which the pieces have to the grain of deal boards. It is sometimes of a chocolate colour, and sometimes of a shining black. The former sort seems to be less impregnated with bitumen, is not so solid and heavy as the latter, and has more the appearance of wood. It lies in straight and even veins, and is frequently dug in pieces of three or four feet long, and, with proper care, might be taken out of a much greater length....When it is first dug, and moist, the thin pieces of it will bend like horn, but when dry, it loses its elasticity, and becomes short and crisp. At all times, it is easily to be separated into very thin

laminae, or splinters, especially if it lies any time exposed to the heat of the sun, which, like the fire, makes it crackle, separate, and fall to pieces.' From Milles' description, it is likely that he was describing actual fossil trunks of the lignite-forming redwood tree *Sequoia*. William Pengelly noted in 1862 that in the uppermost part of Bed 25 in the Bovey Coal Pit (this bed is numbered on illustration 7.3) were slabs of board coal of great length, whose width indicated the presence of trees (probably *Sequoia couttsiae*) fully six feet in diameter.

The burning properties varied with the type of coal. Jeremiah Milles noted that the type which was nearest to the clay burnt heavily and left a large quantity of brownish ash. The wood coal 'is said to make as strong a fire as oaken billets, especially if it be set on edge, so that the fire, as it ascends, may insinuate itself between, and separate the laminae'. But the heat of the stone coal 'is accounted most strong and durable, being apparently more solid and heavy, and probably also more strongly impregnated with bitumen'. When put into a fire it 'burns for some time with a heavy flame, becomes red-hot, and gradually consumes to light white ashes'.

Dr Milles was one of many writers who commented on the foul smell given off when the lignite is burnt. He remarked that: 'The thick heavy smoke, which arises from this coal when burnt, is very fetid and disagreeable....The whole neighbourhood is infected with the stench, which is wafted by the wind to the distance of three or four miles. When burnt in a chimney...the offensiveness is lessened by the draught; however, it is found, by those, who live continually in the smoke of it, not to be unwholesome; nor is it in the least prejudicial to the eyes, like the smoke of wood.' In a footnote to an account of the Bovey Coal by Richard Polwhele in 1797 there is a quote: 'I have actually smelt this coal (says a gentleman of Exeter) more than ten miles off'!

An interesting feature noted by Jeremiah Milles was the presence within the clay, but adhering to beds of lignite, of 'lumps of a bright yellow loam, extremely light, and so saturated with petroleum that they burn like sealing-wax, emitting a very agreeable and aromatic scent'. This material was later described by Charles Hatchett in 1804. He concluded that it was not loam impregnated with petroleum as Dr Milles had supposed, but a substance of 'a peculiar and before unknown nature, being partly in the state of vegetable resin, and partly in that of the bitumen called Asphaltum'; he called it *Retinasphaltum*.

William Pengelly, writing in 1862, noted that spontaneous

combustion commonly occurred in the spoil heaps of waste lignite and clay thrown out from the pit, especially after heavy rain. The fire was not generally visible during the day, but was indicated by smoke and 'the very offensive odour'. Cracks lined with sulphur crossed the burning mass, and occasionally aluminium sulphate crystals were formed. Iron pyrites was common in the spoil tips.

MONTAN WAX

The lignites of the Bovey Basin have not only been used as a fuel, but also considered as a source of organic chemicals, especially a substance called montan wax. This is a type of ester wax, dark brown, hard and odourless in the crude form, which is obtained by solvent extraction from certain types of lignite – it is really fossilized plant wax. It is used today mainly in car and shoe polishes (about 30% of total world production is used in car polish), in electrical insulators, and as a lubricant in the plastics and paper industries. It was formerly used extensively in carbon paper. The earliest commercial production was in Germany in 1903, using Thuringian lignite, and Germany is still the world's largest producer.

There were attempts by a German company before the First World War to produce waxes from the lignites of the Coal Pit, but the results were disappointing, with Aubrey Strahan concluding in 1920 that there was no evidence that the lignite could be profitably extracted as a source of montan wax. It is not clear why the Germans should have been interested in the Bovey lignites when Germany had vast reserves of lignite ('Braunkohle') within its own borders. Perhaps they thought that the Bovey lignite might yield waxes of somewhat different and more desirable character, in view of the different geological mode of formation of the deposit. Nine samples collected by the Geological Survey were analysed and showed that the amount of wax was too small to be worth extracting on a commercial basis.

Despite this, during the Second World War, when the import of German montan wax naturally came to an abrupt end, it was necessary to find a substitute for the wax for 'a specific war purpose', and a detailed assessment by the Geological Survey in 1941-43 indicated that the Bovey lignites contained up to 5% of crude wax soluble in benzene. However, there were thought to be considerable difficulties in the way of commercial production because of the discontinuous nature of the strata and the admixture of the lignite with clay. The investigations were mainly centred in the outcrops along the eastern side of the Basin, especially around Southacre where the 'Big Coal' yielded the richest amount of 5% crude wax. No lignite samples appear to have been examined from the obvious place – the Coal Pit - probably because at that date it was water-filled and no good samples could be obtained.

HISTORY OF THE BOVEY 'COAL PIT' ('BLUE WATERS MINE')

Exploitation of the lignite of the Bovey Basin has always been focussed on the Bovey Tracey area where especially thick beds are developed in the Brimley Member of the Bovey Formation (shown on the map, 6.3, on page 109). There, about a mile south of the town, there was for many years a large open pit, called the 'Coal Pit' or in later years the 'Blue Waters Mine'. Lignite has been dug from the open pit and also underground from tunnels driven out from the bottom of the pit. The site of the pit is now a lake (see photo 7.14), and an estate of park homes has been developed to the south of the pit. There are no longer any exposures of the lignite or clay beds visible in the flooded pit, but fragments of lignite can be found in the waste dumps thrown out from the pit (photo 7.1).

The date of the first working of the lignite from the 'Coal Pit' is uncertain. In 1972 W G Hoskins noted that the lignite of Bovey Tracey was being exploited as early as the reign of Henry VIII, and a Robert Stone was referred to as a 'collyor' [collier - coal miner] in an enrolled deed of 1541. Celia Fiennes, during her celebrated journey through Britain in 1698, mentions Bovey 'coal'.

Lignite in the Bovey Basin has been used as a fuel since the 16th century, despite the unpleasant smell that it gives off during burning. Extensive digging and use of the lignite increased substantially with the establishment of potteries nearby. Peter Weddell and Keith Westcott, writing in 1986, noted that the pottery at Indio was operating in 1766, but there was possibly an earlier pottery of about 1750 in the town of Bovey Tracey itself. The Bovey Pottery was apparently working by 1775; it is not marked on the 1765 map of Devon by Benjamin Donn, which does show the lignite pit and lime kiln. Weddell and Westacott noted that Courtenay family papers in the Devon Record Office (1508M/Special subjects: Mining 25) record that in 1775 money from Lord Courtenay's share of the profits from the lignite mine was used to pay the debts of the pottery. The failures of the pottery were ascribed by the workman to the inadequate nature of the lignite which did not generate enough heat for good firing of the pots.

On the face of it, the clays interbedded with the lignites formed a potential source of pottery clays, but it is surprisingly difficult to establish definitely that the clays were used to any great extent in the neighbouring kilns, or whether they used clay from other areas of the Bovey Basin. Lance Tregoning, writing in 1983, indicates some use of the clay, for he notes that around the end of the 19th century 'some local clay was still being used which was dug from the pit at Bluewaters...', and 'Two [water] wheels were at Bluewaters, one for pumping out the water, and one for lifting out the excavated clay.'

The bottle kilns of the last of the pottery companies, the Bovey Pottery Company Ltd, which closed in 1956, still survive and have been restored (photo 7.2). They now form part of the 'House of Marbles' visitor attraction in Bovey Tracey.

One of the earliest and most detailed accounts of the Bovey Coal Pit is that of Dr Jeremiah Milles in 1760 and I have described the varieties of lignite recognised by him above (page **138**). The lignite deposit was clearly being actively dug at that date, for Milles noted the presence of 70ft of clay and lignite in an open pit. In the upper part, the lignite was in beds from 18 inches to four feet thick, separated by beds of brown clay; the thickness of the lignite beds increased with depth, and the lowest bed was 16ft thick. It rested on a bed of clay under which was a bed of 'sharp green sand, not unlike sea sand, 17ft thick'. He noted that 'From the sand arises a spring of clear blue water, which the miners call mundic water' (this may be the origin of the name 'Blue Waters Mine' which was later applied to the pit in the mid-20th century). In 1862 William Pengelly and Oswald Heer (see below) measured a total of 125ft of strata in the Bovey Coal Pit, including a further series of lignites present below the bed of sand referred to by Milles, which were not known in 1760, but which were probably discovered in 1797.

Jeremiah Milles had strong opinions on a question which to us today seems to have an obvious answer: he believed that the lignite was not derived from plants, but was mineral in origin. Against him stood a Professor Hollman of Göttingen, Germany, who maintained that the lignite was vegetable in origin. Surprisingly, this rather fruitless controversy between the two stubborn men was maintained for over 20 years! Milles noted that the 'coal' was dug from an open pit, together with the interbedded clays, and 'though it lies very close and compact in its original bed, yet it is so easily separated, that they can afford to sell it for half a crown a ton at the pit. The

7.2. The bottle kilns of the former Bovey Pottery still survive and now form part of the 'House of Marbles' visitor attraction in Bovey Tracey.

smaller coal is separated from the clay by a screen, or grated shovel; the larger, which rises sometimes in pieces of above an hundred weight, is piled up by hand. There is hardly any other use made of it at present, but to bake the earthenware of a manufacture erected at South Bovey, and for burning of limestone, which rising in great quantities at the neighbouring town of Chudleigh, the coal is carried thither, and they return with limestone to the pit, which they burn there, for the use of the Northern parishes, to whom it lies more convenient than the kilns of Chudleigh'. Richard Polwhele's description of the Bovey Coal in his 1797 *History of Devon* was apparently largely based on Dr Jeremiah Milles' 1760 account.

The earliest known financial accounts of the 'Coal Works and Lime Trade at Bovey Tracey' between 1756 and 1764 can be found in the Devon Record Office (1508/Devon/Mining/25). The profit and loss figures show a gradually improving financial situation during the period, as follows: 1756 - loss of £160; 1757 - loss of £105; 1758 - loss of £70; 1759 - profit of £42; 1760 - profit of £164; 1761 - profit of £269; 1762 - profit of £358; 1763 - profit of £422; and 1764 - profit of £462.

Dr Maton, in his *Observations on the Western Counties of England*, in 1794 and 1796, noted that the Coal Pit was in use to serve the pottery close by, but the pit was often filled to a height of 40ft or more with water which was pumped out by machinery. By 1800, tunnels had been pushed out from the bottom of the pit, which was drained using a 24ft diameter overshot water wheel. Robert Scammel, writing in Parkinson's *Organic Remains* of 1804, gave a section of a shaft 75ft deep, and numbered the lignite beds 1 to 17 from top to bottom. Bed 15 was called the 'great bed' and was 5ft 6in thick in four 'floors'; just beneath it was bed 16, the 'little bed' and the lowest bed, (No 17), 2ft 7in thick, was called the 'last bed'. He wrote that 'The Bovey Coal is now [1804] used for supplying the steam-engine, for burning lime; and, occasionally, for giving the earthen-ware its first burning; it is not now used for domestic purposes, the sulphurous gas it emits being, not only extremely disagreeable, but injurious to the health of the inhabitants'. He noted that the workmen recognised three types of lignite: knotty coal, stone coal and board coal.

Jean De Luc, who we have met earlier in Chapter 4 when he gave an account of the whetstones of the Blackdown Hills, also wrote about the Bovey Coal in his *Geological Travels* of 1811 (based on a journey made in 1806). Although he noted 'The common opinion respecting the combustible substance here worked... that it consists of wood', he doubted this on various grounds and thought instead that the 'brown strata...consist only of *clay*, impregnated with *bitumen*'.

An 1837 hand-drawn sketch-map titled *Map of part of Bovey-Heath and of several watercourses belonging to the Folly Pottery* (Devon Record Office 1508M) shows a 'Coal Works' south of the Folly Pottery, with an 'Engine House' and 'Whim' marked (a whim is a capstan or drum with a vertical axle which was used for hauling material from the pit to the surface). A 'Tram Road' marked on the map between the Pottery and the lignite pit is clearly the Haytor Granite Tramway (see page 145). Two limekilns are shown close to the south side of the tram road and near the lignite pits.

In White's *History, Gazetteer and Directory* of 1850 the Bovey Coal was noted as being used 'by the poor' and also at the extensive pottery, established in 1772, where 300 people were employed. The manager of the Bovey Tracey Pottery Co was William Robinson and the sub-manager was William Sharland. In 1893 *Kelly's Directory* noted that: 'Lignite or "Bovey Coal" has been worked here for more than a century by means of deep cutting and tunnels; it is used chiefly in the neighbouring pottery, its sulphurous smell rendering it unfit for domestic use'. It was also noted that 'Many persons are also employed at Mr G R Divett's potteries and lignite works. A special kind of pottery called 'Chudleigh Ware' is made here'. The sole proprietor of the Bovey Tracey Pottery Co was G R (George Ross) Divett.

As we shall see below, there was a short-lived attempt to exploit the lignite for fuel just after the Second World War, in 1945-49. There have also been several schemes to extract montan wax from the lignites, beginning with a German Company which operated the Coal Pit just before the First World War, but none has been successful.

The 'Coal Pit' in 1859-60; William Pengelly and Oswald Heer

The 1862 work of William Pengelly and Oswald Heer (issued in 1863 as a separate volume entitled *The lignite formation of Bovey Tracey*), marked the beginning of a more scientific approach to the study of the lignites and their plant fossils. William Pengelly made accurate measurements of three vertical sections in the 'Coal Pit', the thickest of which showed 125ft of lignites and clays with rare sands. Two of his sections are illustrated in 7.3 and 7.4. The strata are not

7.3. *A cross-section through the lignitic strata of the Bovey Coal Pit, based on measurements made by William Pengelly and Oswald Heer in 1860. The dip of the beds is 12½° to the SSW. The beds from which plant fossils were collected are numbered. Reproduced from Plate II of Pengelly and Heer (1863).* Courtesy of Special Collections, University of Exeter.

identical across the pit: the beds down to a depth of 80ft 8in in the pit were measured again at a point 680ft farther east and showed a thinning out of about 20%. William Pengelly noted that the lignite beds weathered out in relief on the wall of the pit, like moulding, so that it was possible to make out in a rough way the succession of the beds. However, the details were obscured by clay and sand washed down the sides of the pit, so it was decided that to make accurate measurements and collect fossil plants it would be necessary to cut a series of steps on a large scale from top to bottom of the pit. One wonders what the local people must have made of the sight of this gentleman geologist and his companions diligently measuring and collecting their way down a series of steps cut in the pit sides. There must have been much speculation and maybe suspicion in the local cottages.

The fossil plants collected from the Coal Pit were examined and identified by the Revd Oswald Heer, Professor of Botany in Zurich, who described fifty species of plants and referred them confidently to the period of earth time called the Lower Miocene (see table 1.2).

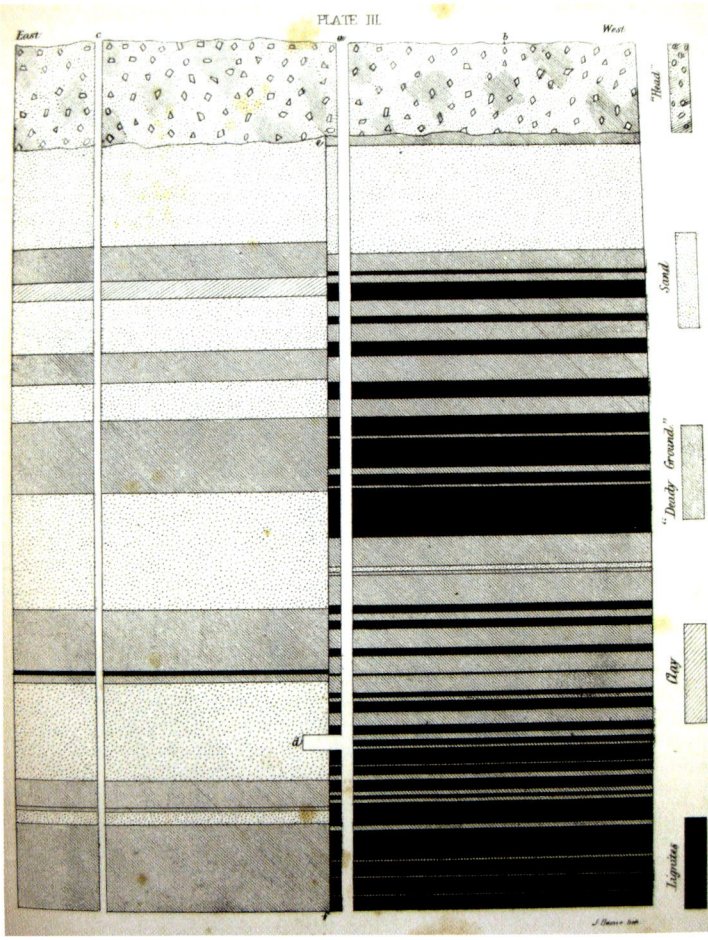

7.4. *Plate III from William Pengelly and Oswald Heer's 1863 description of the Bovey Coal Pit, showing a fault (marked e-f) located not far east of the Pit, but not seen in it (a fault is a dislocation along which relative movement of the strata on either side has taken place). To the east of the fault, the strata are mainly clay and sand; to the west they are mainly lignite and clay. The total depth of the section is 99ft. (a) is the position of the old 'engine shaft'; (b) is the eastern end of the Coal Pit, 336ft west of (a); (c) is the position of a borehole sunk by Mr Divett, 420ft east of (a); (d) 80ft below the surface, is a horizontal tunnel which crosses the fault from the lignite sequence into sand.* Courtesy of Special Collections, University of Exeter.

Above left: **7.5.** *A reproduction of Plate XVI, an example of one of several plates in Pengelly and Heer's 1863 work on the Bovey lignite. The fossil leaves illustrated belong mainly to* Cinnamomum, *a genus of evergreen trees and shrubs which is included in the Laurel family. 1-8:* Cinnamomum lanceolatum; *9-16:* Cinnamomum scheuchzeri; *17, 18:* Cinnamomum rossmässleri: *19:* Sclerotium cinnamomi. *(These names may have been superseded since 1863).* Courtesy of Special Collections, University of Exeter.

Above right: **7.6.** *Details of* Sequoia couttsiae *which makes up the bulk of the Bovey lignites. Reproduced from Plate VIII of Pengelly and Heer's 1863 work on the Bovey lignite.* Courtesy of Special Collections, University of Exeter.

Examples of plates from the beautifully illustrated descriptions of the fossil plants are shown in 7.5 and 7.6. The presence of iron pyrites in the lignite, and the speed with which it dries out, make the plant fossils from the lignite very difficult to preserve. William Pengelly soaked the plant specimens in gelatine which slowed down their decay. I am uncertain whether his plant collection has survived, but I suspect that the specimens would have all decayed in the 150 years since they were collected. Much later work by Marjorie Chandler in 1957 showed that many of Heer's determinations were unsatisfactory by modern standards. She concluded that, based on material from the Bovey 'Coal Pit', and a 'pit at Kingsteignton' an Oligocene and perhaps Middle Oligocene age was indicated, but she did not rule out the possibility that the beds below the Coal Pit level were Eocene in age and the beds above could be Miocene.

William Pengelly records some interesting information about the underground workings in about 1859 or earlier. He noted that: 'Subterranean excavations have been carried on very extensively, in various directions, by means of tunnels opening out of the pit at its bottom. At present the working is confined to one tunnel, extending 190 fathoms [1140ft], almost in a straight line, in the direction N. 65° W. [295°] from the western end of the pit." He says that 'At some parts of the workings the lignite is so firm, compact and tough that it can only be removed by blasting; at others it is extracted without difficulty, or is even loose and brittle. Occasionally, though apparently of great firmness, large slabs fall from the top (locally "back") and sides of the tunnels, not, however without giving timely warning in the shape of a smart crackling noise'. The air in the innermost 90 fathoms [540ft] of the working was 'oppressively foul' and the workmen had difficulty in keeping their candles burning. The temperature was 61° F in a dry part of the inner end of the tunnel. All in all, conditions were clearly quite unpleasant at the western end of this long tunnel. The amount of water in the tunnels was quite variable. In the western tunnel, mentioned above, Pengelly noted that: '....the drip of water is very considerable, amounting, indeed, in a few instances to a continuous and rather copious flow......In other parts the same tunnel is quite dry...'

In 1859 the lignite in one of the old tunnels had been set on fire (possibly by burning cinders from a nearby brick kiln), and could not be put out. The owner, Mr Divett, decided that the only way to extinguish it was to stop pumping and allow the pit to flood, which was done. Thus, when in the spring of 1860 the celebrated geologists Sir Charles Lyell and Dr Hugh Falconer visited the area, they found the water in the pit to be 30ft deep, and Mr Divett had no intention of pumping it out. Fortunately for the study planned by William Pengelly, described above, Mr Divett, perhaps influenced by the influential visitors such as Lyell '....changed his mind, soon after we began our investigations; and as we descended with our section, we had the gratification of seeing the water steadily retreating before us, so that the lowest beds were exposed some time before we reached them'.

Workings east of the pit are indicated by the record in Pengelly's 1862 account of a shaft (the 'old engine shaft'), shown on 7.4, mentioned as being 56 fathoms (336ft) east of the pit, but little is known about them. The shaft was located close to a fault which separates mainly clay and sand to the east from mainly clay and lignite to the west. Workings driven eastwards from the shaft crossed the fault, and the proprietor, Mr Divett, recorded that 'I drove towards it [the fault] in many places, and always found the 'coal' fail and replaced by hard and wet 'deady' clay. At one place I drove further and cut into a bank of sand full of water, which ran into the shaft and 'starved' the pump for some time' - (see (d) on 7.4).

There is little available information about the working methods at this period (the mid-19th century). Presumably they were rather like coal mining, with the use of picks and shovels to dig out the lignite which was then loaded into tubs which probably ran on rails. We know from William Pengelly's description that blasting was used at times to remove the hardest lignites (the 'stone coals' of Jeremiah Milles). It is uncertain whether the tubs were pulled by horses, or whether man-power was used. The roofs of the tunnels were mainly supported by pillars of lignite left in place, although presumably timber supports were used where appropriate. The pit was pumped dry of water by a large water-wheel, which was also used to raise the lignite and waste material to the surface by an inclined plane at the eastern end of the pit.

There is an interesting commentary by William Pengelly on the use of lignite as a fuel in the Bovey Pottery in 1859-60. On account of the flooding of the pit, mentioned above, he says that ordinary 'sea-borne' coal was then used exclusively at the pottery, and little lignite was used. The price of lignite was 5 shillings per ton, but was not much in demand owing to its offensive odour. The calorific value of the lignite was low, and it was estimated that about six tons of lignite produced the heat of one ton of ordinary coal. Until 1853 there are no reliable figures for the amount of lignite produced. The 1871 Coal Commission report gave the following figures for output, showing the decline in production: 1853 – 18,633 tons; 1856 – 3,850 tons; 1857 – 5,660 tons; 1867 – 1,368 tons; and 1868 – 1,383 tons.

With the extension of the railway network to many parts of the country, and to Bovey Tracey in 1866, coal was readily available just about everywhere at a reasonable price, and its advantages over lignite (lack of offensive odour and superior calorific value) meant that the use of lignite as a fuel for the Bovey Tracey potteries fell into disuse, although there may still have been a small production for local domestic use for those who could not afford coal. Woodward (in a manuscript note) wrote that by 1875 the use of the lignite had ceased. However *Kelly's Directory* for 1919 noted that the lignite was worked up until 1894. The pit lay derelict and presumably flooded until the arrival of a German company in 1914 (see below). Ordnance Survey 6-inch-scale maps ranging in date from 1890 to 1930 (for example, that of 1890, 7.7) show very little change in the outline of the flooded Coal Pit, which had an area of just over 3 acres.

The 6-inch-scale Ordnance Survey map of 1890, reproduced in illustration 7.7, shows a 'Lignite Shaft', (with adjacent engine house and smithy), about 260ft west of Ashburton Road; it is possible that this may link with the 1140ft-long tunnel noted by William Pengelly (see page 144) as extending west from the Coal Pit on a bearing of 295°. The site of these structures is still an open field (in 2010), but there is no trace in it of any remains of the lignite industry. The same map shows a 'Lignite Works' just northeast of Pottery Cross. Another 'Lignite Shaft' is marked on the map between the Lignite Works and the pit. A building marked south of Pottery Pond may be another lignite works but it is not identified on the map. However, a tramway (not the Haytor Granite Tramway, described below) connects it to the pottery farther east. The sites of these shafts and works are now occupied by houses.

Also visible on the map (7.7) is part of the course of the celebrated Haytor Granite Tramway which is unique because the 'rails' were made from solid blocks of granite cut to form flanges to guide the trams. It is marked just south of the words 'Vagabond Stone' on the map and from there extends east-south-eastwards to pass immediately north of the former Lignite Works near Pottery Cross, and from there to just south of Pottery Pond. The tramway was constructed by George Templer to carry granite from quarries at Haytor on Dartmoor, to a canal basin on the Stover Canal at Ventiford. It was opened on 16 September 1820 to great celebration and was then 7 miles long (it was later extended to 9 or 10 miles). The vertical drop from Haytor to Ventiford is about 1300ft, and the tramway had to follow a tortuous course to overcome the steep gradients. The empty trams were pulled upwards to the quarries by teams of horses; the loaded trams ran downhill by gravity to the canal basin. However, the enterprise had a life of less than 40 years, and by 1858 the quarries on Dartmoor were deserted and the tramway had fallen into disuse.

Much of the course of the former tramway is now followed by a footpath called the Templer Way, named after the builder of the tramway. Some details may be of interest for those who wish to follow it in the immediate vicinity of the old lignite works. Westwards from Pottery Pond, the tramway crosses two roads (first Ashburton Road, then Brimley Road) before reaching Stentiford Lane about ¾ mile from Pottery Pond. The section from the Pond to the Brimley Road crossing is straight and lies between gardens. About 130 yards west of the southern end of Pottery Pond, the old milestone 3 can still be seen on the north side of the tramway (photo 7.8); distances were measured from the canal basin at Ventiford. There are no traces of granite rails between Pottery Pond and Brimley Road, but in 1964 M C Ewans noted that near milestone 3 the rails could then still just be seen showing through a surface of ashes forming the path, and that the junction with the road near Pottery Pond was the last time they could be seen as a continuous whole. It is uncertain whether the granite rails are still there but buried, or have been removed. Going westwards again from where the tramway crosses Brimley Road until it meets Stentiford Lane there are several places where the flanged granite rails can still be seen (photo 7.9). There are several gentle curves along this stretch. The gauge between flanges is about 4ft 3in, and the flanges are generally about 3in deep.

The location of the Haytor Granite Tramway suggests that it could have been a convenient way of transporting lignite from the works near Pottery Cross eastwards to the potteries, but there is no definite evidence that it was ever used in this way. The lignite works shown on the 1890 Ordnance Survey map (7.7) is laid out at right angles to the tramway, and its northern end extends over the course of the tramway, hinting at some connection between the two. We might speculate that there was a loading bay for lignite to be transferred from the works into tubs running along the tramway.

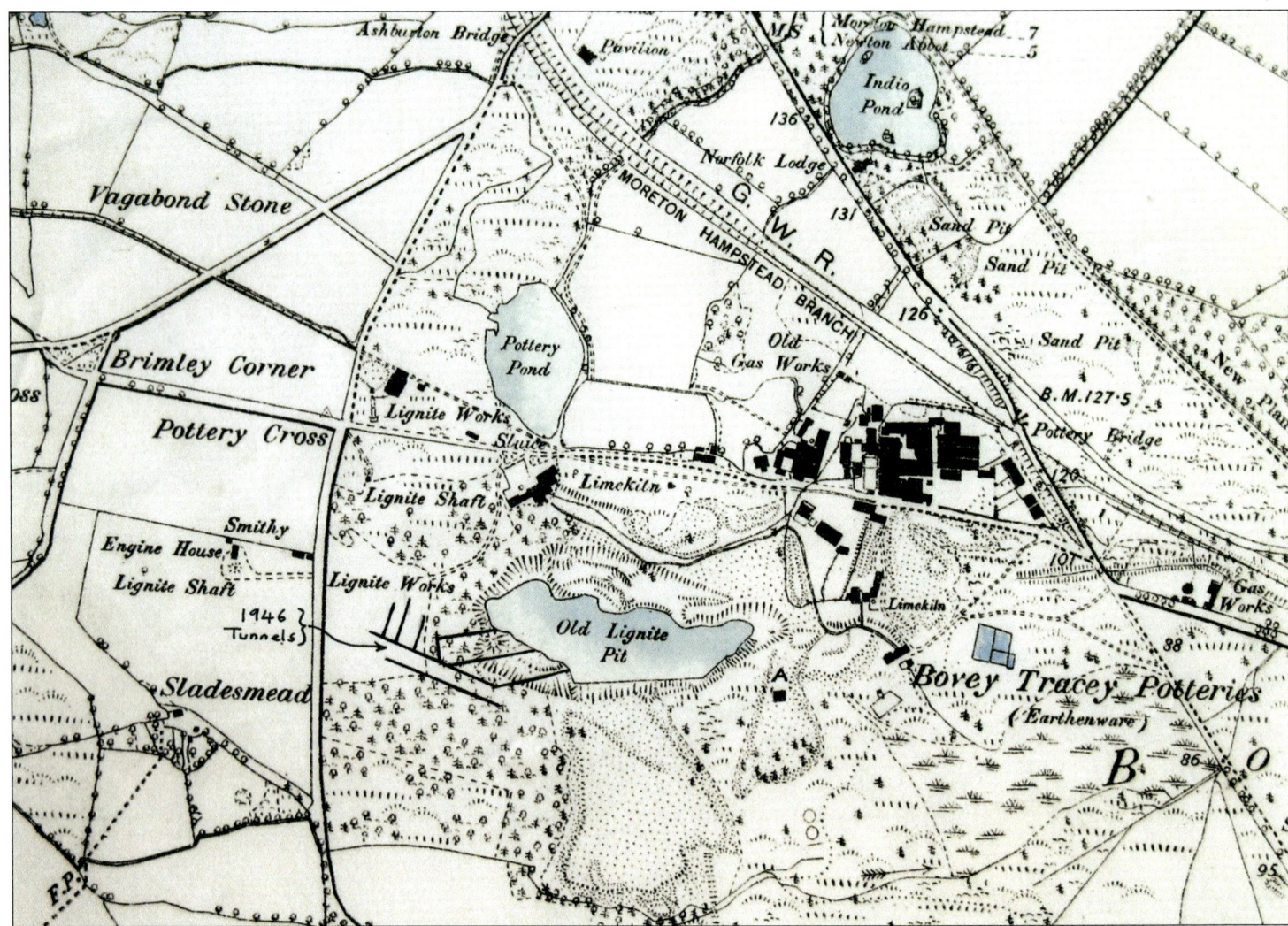

7.7. *An extract from a 6-inch-scale (1:10,560) Ordnance Survey map of 1890 showing the extent of the Bovey 'Coal Pit' (marked 'Old Lignite Pit') at that date and the locations of shafts, lignite works, tramways etc. This map also shows the approximate extent of the 1946 tunnels west of the Bovey Coal Pit, superimposed on the 1890 base map. The size of the pit changed very little between 1890 and 1945. Inclines were driven obliquely down the dip and headings along the strike of the beds, as shown. The position of the 'German Electric Works' (see page 148) is marked by the letter A to the east of the pit. Not to original scale.*

7.8. Milestone 3 on the Haytor Granite Tramway near Pottery Pond, Bovey Tracey.

7.9. The remains of the granite 'rails' on the Haytor Granite Tramway between Brimley Road and Stentiford Lane, Bovey Tracey.

However we do know that there was another tramway (shown on the map 7.7) connecting a works just south of Pottery Pond to the pottery farther east and also branching to serve a limekiln.

The Germans and Montan Wax

According to the 1918 account by Aubrey Strahan and others, dealing with lignite resources, a company called the Ilsington Mining Company had 'some years previous to 1914' acquired an area of 26 acres which included the old 'Coal Pit' and a further 110 acres, in order to exploit the lignite, it seems with the main aim of producing bitumen. Apparently a small installation in which gas was produced was built, and dynamos were attached to this for the production of electricity for the works. Now we come to the first mention of a German company. The Ilsington Company found it difficult to produce a guaranteed percentage of bitumen but 'A method of extracting the bitumen and deriving a profit from their holdings was shown to them by the German Mining Co'. The German firm was interested in the possibilities of lignite not as a fuel but as a chemical raw material, particularly for the production of montan wax.

It is not clear what the relationship was of the Ilsington Mining Company to the 'German Company'. However, it seems that in March 1914, the German company began drilling boreholes, and by the end of May 1914 three had been completed and two were in progress. The Germans left abruptly in 1914, two days before the declaration of war. The works were visited by Mr Pringle of the Geological Survey in 1917; he found that the engine house and other buildings were intact, and even a derrick for one of the four boreholes sunk by the German company to establish lignite reserves was still in place.

The German company used the lignite to generate electricity. The power station was about 100 horsepower, and two vertical engines driving generators were worked from gas made from lignite by burning it in a producer. The electricity was carried by aerial wires 1¼ miles to Liverton for lighting the village. The probable location of the German electricity works east of the open pit is marked by the letter A on the map, 7.7.

Between the wars

Amery Adams noted in 1946 that 'About 1905 a project was seriously considered to electrify the old railway from Bovey to Haytor [this was the Haytor Granite Tramway – see page 145], generating the electricity required by using Bovey coal to make producer gas. The scheme went so far that a small power station was erected near the potteries and electricity made as planned was used for some industrial (pottery) purposes. The ruins of this station are still there [in 1946]. Mercifully for the promoters of the idea, and I think I may say for all lovers of Dartmoor, the electric tramway scheme did not develop as by another 10 years the development of the motor car would have killed it'.

A report in *The Times* newspaper of 3 February 1920 referred to a £10 million scheme to establish a copper refinery fuelled by Bovey lignite. A guide to Newton Abbot, published in 1920, described this somewhat unlikely sounding plan, in which it was stated that 'It is anticipated that electricity will be produced in bulk…from the lignite deposits estimated at many millions of tons…so that subsidiary industries may be opened up to the advantage of the town

and district…a Syndicate has been formed for the purpose of opening up on the banks of the River Teign an electrolytic refinery for the refining of copper smelted abroad'. I have not been able to found more about this scheme, which clearly came to nothing.

The 1945-49 re-opening of the 'Coal Pit' – the 'Blue Waters Mine'

Immediately after the Second World War, Britain was in an exhausted state. Fuel supplies were in short supply, and the winter of 1946-47 was one of the hardest on record. The fuel crisis prompted people to look at alternative sources of fuel, and attention focussed on the lignite deposits of Bovey Tracey as a way of helping out. For a brief period, the area became a hive of activity: the old 'Coal Pit' was drained, and new drives were pushed out from the base of the pit (shown on the map, 7.7). This initiative was the brain child of Mr J H Wilson, who had a long-term interest in the Bovey deposits, particularly for the production of montan wax, and acted as technical advisor. However, the fuel shortage stimulated the idea of using the lignite as a fuel, and a company called British Lignite Products Ltd was set up, under the chairmanship of Mr C W Parish. Interest in the project was considerable, and the 300,000 Ordinary shares at 3/6d were apparently oversubscribed, although some of the subscribers may have later come to rue their enthusiasm! The authorised capital of the company was £500,000. The company was apparently a subsidiary of a Mexican company called the El Oro Mining and Exploration Co which probably provided finance. A *Times* report of October 22 1946 stated that the El Oro Mining and Railway Co had liquidated its Mexican assets and was placed in voluntary liquidation, and a new company formed – the El Oro Mining and Exploration Company. An interest in the lignite deposits of Bovey Tracey was acquired by the company in May 1945. The *Times* of May 23 1947 reported that shareholders of the El Oro Mining and Exploration Co were offered 222,222 3s 6d shares at 4s 6d each in the ratio of 40 for every 207 held. The proceeds were to be used to subscribe for future shares in British Lignite Products. El Oro proposed to increase its already large shareholding in the new company so that it would eventually control it. The issued capital of El Oro was then £287,000 and the shares were quoted on May 22 1947 at 5s 6d.

The degree of involvement of the Government with the project is not absolutely clear. Mr Parish, in his 1947 booklet called *The Creation of an Industry* which was very much a 'puff' for the company, remarked that he had met Sir Stafford Cripps, President of the Board of Trade, in January 1947 and he was quick to see the importance of the project. Since then the Board of Trade had been of help and the Ministry of Fuel and Power (after initial hesitation) had helped since the summer of 1945.

The Times of June 23 1947 carried an optimistic report on the prospects for British Lignite Products. The optimism centred on montan wax, which was stated to be scarce in other lignite deposits of the world. Because of its special use in insulation it commanded a high price - £300 per ton. It was reported in *The Times* article that exports of montan wax to Canada and the USA were expected to reach an annual value of $2,000,000 when the plant at Bovey Tracey, then being built, was in full production. The plant was expected to cost £500,000, but it would be some time before it was complete and 9 months before montan wax could be produced on a commercial scale. As far as I know the plant was never completed, and I believe that no montan wax was ever exported. Mention was also made of the production of lignite briquettes to be sold as fuel. The target outputs were given as 3,000 tons of montan wax, and 500,000 tons of lignite briquettes per year. There were ambitious plans to build a miner's village, but in the meantime, the Devon House of Mercy in Bovey Tracey was converted into flats to accommodate the lignite miners, many of whom were from South Wales. P H W Scott notes that 200 men were employed.

Drainage of the Blue Waters Mine pit began in July 1945, and 864,000 gallons of water a day were pumped for 10 weeks until the middle of September, by which time the pit was essentially dry (photo 7.10). When the pit had been de-watered, the 1860 workings were found to be still standing intact, but falls began after the exposure to air, followed by temporary flooding due to overflow of water from a leat on the north side of the quarry. Driving then began through the old workings. Inclines were driven obliquely down the dip of the strata (which is about 15 to 20° to the south-south-east), and then headings along the strike (as shown on the map, 7.7). The lignite bed was reported as being 30ft thick. The drives and stalls in the lignite were about 10 to 15ft high (photo 7.12), and kept within the thickness of the lignite, otherwise the clay roof would be prone to falls.

Concurrent with the beginning of lignite extraction, exploration for further reserves began in 1945 with the sinking of several boreholes (El Oro boreholes numbers 1 to 4). The boreholes sunk by the German Company in 1914, mentioned earlier in this chapter, had already provided valuable information. Evaluation of the boreholes suggested lignite reserves of 53 million tons.

The unusual fact of lignite mining in England stimulated quite a lot of press interest at the time, and there were several articles in various papers. For example, an article in the *Manchester Guardian* of November 12 1947 noted that at first 100 tons of lignite a day had been extracted by deep mining, but since July (presumably of 1947)

7.10. A view of Blue Waters Mine in September 1945, after de-watering.

7.11. Blue Waters Mine in 1947, showing mine cars near the mine entrance.

the output had been increased to 600 tons a day by opencast methods. Already the men had excavated 80ft, uncovering a seam of lignite nearly thirty feet thick, and ultimately they would reach the old underground workings. For four months the weather had been kind to them, and whatever it did then, they would be able to work through the winter. The photo (7.11) shows empty mine cars descending to the mine entrance in 1947.

Such were their high hopes for the long-term potential of the Bovey Basin that the directors of British Lignite Products Ltd commissioned a report on 11 April 1947, by a Welsh consulting engineer, W Brynmor Davies, on the advisability and practicality of opening up a new 'coalfield' on a comparatively large scale. The report was optimistic, and 400 million tons of lignite were estimated to be present in the basin, but we know today, with our improved understanding of the Bovey Basin, that its conclusions were to a certain extent based on insufficient knowledge of the structure and geology of the basin. Opencast mining and extensive underground mining were proposed, with the establishment of four collieries with slants driven down from surface along the seams. Each colliery was to produce 250,000 tons of lignite per year. This scheme pre-supposed a regular and continuous outcrop of the lignite beds all around the basin which is not in fact the case.

Digging the lignite 1945-49

The above-ground operations in the Blue Waters Mine were quite modern and mechanised, for the *Manchester Guardian* article of 12 November 1947, mentioned above, says that the scene in the pit was like an opencast coal mine, with tractor-drawn scrapers, dragline excavators, diesel-powered shovels and bulldozers working.

The problems of underground mining of the lignite were looked at by J Norman Wynne in a *Mining Magazine* article of December 1947. He noted that after exposure to air, the lignite rapidly crumbled and broke up, consequently weakening the sides and roof of the tunnels, Square-setting would be too costly, and sand-filling not practical because of the dampness of the lignite through which sand-passes would have to be cut and filling poured. Using the pillar-and-stall method of mining would require leaving unusually large pillars because of the continual crumbling of the drying surface of the lignite, made worse by the concussive effects of blasting. The result would mean that in the end only a limited proportion of the lignite seam could be safely extracted. Taking all this into account,

7.12. Blue Waters Mine in 1947. A working face in a drift shows a thick bed of lignite, with thin beds of pale clay. The workings are partly supported by steel arches.

7.13. Blue Waters Mine in 1947. The picture shows miners operating a Siskol coal cutter which was obtained second-hand from a Scottish coal mine.

it was decided by the general manager, Mr Varvill, to excavate by opencast methods, but to continue the drifts and cross-cuts for the purpose of exploration and drainage, albeit of reduced dimensions. These tunnels, 8ft by 8ft in cross-section, were supported by steel arches with timber lagging where necessary (photo 7.12 shows steel arches in use).

As for the actual methods of cutting the lignite underground, J Norman Wynne remarked that it was not easy to cut lignite by mechanical pick or spade, because the friction quickly heated the cutting tool and caused the lignite immediately around it to swell and grip the metal of the tool tightly. The method used was therefore to drill into the lignite bed with an auger-bit drill, and then only a few lightly-loaded shot holes were needed to loosen the lignite ready for loading onto side-tipping tubs, using hand-pick and shovel. Following this method the lignite face was advanced by about 3ft per shift, an output of 7 tons. Three second-hand 'Siskol' undercut coal cutters (pictured in use in photo 7.13) were acquired from a Scottish coal mine and no doubt improved output.

On the surface, Le Tourneau caterpillar-drawn scrapers, with a capacity of 10 to 12 cubic yards, crowding shovels (capacity of 1 cubic yard) and a dragline excavator (capacity 1¼ cubic yard) were used to excavate the exposed lignite beds. Opencast working started only in July 1947, but an exceptionally dry spell of weather meant that operations could be continued day-and-night. By the end of October 1947 monthly production had reached 15,000 tons. The lignite was sold 'as mined' after crushing and screening. The entire output from the mine was marketed by Renwick, Wilton and Dobson Ltd, of Torquay. John Fox, writing in December 1948, noted that the current capacity of the mine was about 500 tons per day, and the current prices at the mine sidings were: *Cobbles*: 41 shillings per ton; *Nuts*: 39 shillings per ton; and *Smalls*: 30 shillings per ton. He noted that a briquetting plant was being erected and it was hoped to produce 'ovoids' (an egg-sized product made from a mixture of pulverised, washed and dried lignite and coke dust), to be sold under the trade name 'Lignuts'. About 1,820 tons of Lignuts had been sold by the end of the financial year on March 31 1949. The post-Second World War opencast and mining operations considerably increased the size of the open pit and its western edge now extended almost to the Ashburton Road. The adjacent spoil tips roughly doubled in area.

The end of lignite mining at Blue Waters Mine

By the end of the financial year on March 31 1949, the company had lost £26,950. Such losses could not be sustained, and sadly, the scheme, which had started with such high hopes, foundered and British Lignite Products Ltd went into liquidation. Fuel supplies within the UK had improved and the montan wax scheme came to nothing when Imperial Chemical Industries developed a substitute which sold at £150 per ton, a price which was considerably less than that of montan wax produced from Bovey lignite. The plant which had been constructed to produce ovoids was bought by Watts Blake Bearne, the ball clay producers, but it was never used.

The pit filled with water and lay disused for many years; I recall visiting it in the late 1960s and walking on the spoil heaps of lignite and clay waste; the pit was flooded and little was to be seen of the strata formerly so well exposed in the sides of the pit. The Exeter *Express and Echo* newspaper reported on May 20 1982 that the Exeter and District Angling Association had acquired the Blue Waters Lake. Work had already started to reshape some of the high banks around the pond to provide a level causeway and give anglers a convenient height from the water to fish from. The water was known to contain tench, with an 8lb specimen recorded. There were plenty of good perch and thousands of rudd. Some big eels were in residence in the deep water.

The old waste tips have been removed, and the land to the south of the Blue Waters Lake developed for park-style housing; there is currently no public access to the lake and its environs.

Large reserves of lignite remain in the Bovey Basin, particularly around Bovey Tracey. The lignite remains a potential fuel for power stations and other possible uses include gasification (J Norman Wynne, writing in 1947, gave a reported calorific value of about 450 British Thermal Units per cubic foot for town gas produced from Bovey lignite). Lignite has been used extensively in other countries, for example Germany, to produce synthetic liquid fuel. However, environmental considerations suggest that the post-war boom was the last gasp of large-scale lignite mining in Devon and the Blue Waters Mine is most unlikely ever to open again as a lignite pit. Today the pit is flooded and is now a peaceful scene (photo 7.14) compared to the feverish activity 65 years ago when men toiled to dig out the local form of 'black gold'.

7.14. *A view of Blue Waters Lake in 2009, looking to the east.*

SOURCES

Chapter 1. Introduction

Ogg, J G, Ogg, G and Gradstein, F M. 2008. *The Concise Geological Time Scale*. (Cambridge: Cambridge University Press).

Sedgwick, A and Murchison, R I. 1839. Classification of the older stratified rocks of Devonshire and Cornwall. *London and Edinburgh Philosophical Magazine and Journal of Science*, Volume 14, pages 241-260, and page 354.

Chapter 2. Echoing Chambers: The Penn Recca Slate Mine, Buckfastleigh, South Devon

Anon. 1992. Penrecca mine, Buckfastleigh. *William Pengelly Cave Studies Trust*, Volume 64, page 27.

Born, A. 1988. Blue slate quarrying in south Devon: an ancient industry. *Industrial Archaeology Review*, Volume 11, No. 1, Autumn 1988, pages 51-67.

Bowman, A and Tilley, C. 1997. The quarrying industry. Pages 103-111 in *Exmoor's Industrial Archaeology*, edited by M Atkinson. [See pages 104-107 for the slate industry, especially Treborough]. (Dulverton: Exmoor Books).

Hemery, E. 1982. *Historic Dart*. (Newton Abbot: David & Charles).

Hoskins, W G. 1972. *Devon*. (Newton Abbot: David & Charles).

Kelly's Directory of Devon, 1902, page 644 for note on 'Pen Recca'.

Pearman, H. 2003. Penn Rica [sic] Slate Mine, Devon, Saturday, 27th September. *Kent Underground Research Group Newsletter* 79, December 2003, pages 3-4 [Includes a sketch plan of the Lower Series of passages at Penn Recca].

Proctor, C J. 1984. Stemple Passage, Penn Recca Slate Mine. *Devon Speleological Society Journal* (128), July 1984, pages 10-11.

Reed, E and Reed, J. 1989. Early days at Penn Recca – some excerpts from the HQ log [January 29th 1949; January 30th 1949; February 2nd 1949; February 6th 1949]. *Devon Speleological Society Journal* (141), July 1989, pages 6-8.

Reid, C, Barrow, G, Sherlock, R L, MacAlister, D A, Dewey, H and Bromehead, C N. 1912. The geology of Dartmoor. *Memoirs of the Geological Survey of England and Wales. Explanation of Sheet 338*. (London: HMSO).

Shaw, T R. 1952. Penn Recca Slate Mine, Staverton, Devon. *Cave Science (British Speleological Association)*, Volume 3, No 21, pages 199- 222.

Staverton's history. http://www.staverton.com/history/history.htm

Ussher, W A E. 1912. The geology of the country around Ivybridge and Modbury. *Memoirs of the Geological Survey of England and Wales. Explanation of Sheet 349*. (London: HMSO).

Chapter 3. Coal and Paint: The 'Culm' Mines of North Devon

Acworth, R. 1991. The anthracite seams of north Devon. *Journal of the Trevithick Society*, Volume 18, pages 117-125.

Anon. 1933. The production of Bideford Mineral Black. A description of the works of Bideford Black Limited, near Bideford. *Industrial Chemist*, February 1933, pages 45-48.

Anon. 1996. *Archaeological assessment of the DCC Bideford East-the-Water industrial link road*. Exeter Archaeology, Report No 96.71, December 1996.

Arber, E A N. 1904. The fossil flora of the Culm Measures of North-west Devon, and the palaeobotanical evidence with regard to the age of the beds. *Proceedings of the Royal Society*, Volume 74, pages 95-99.

Arber, E A N. 1911. *The coast scenery of north Devon*. (London: Dent).

British Geological Survey. Various files and letters relating to culm, formerly held at the office in Exeter.

Brooke Index. List of mines held at the Westcountry Studies Library, Exeter.

Chope, R Pearce. 1902. Mining at Hartland. Notes of the past No 45. *Hartland Chronicle*, No 73, page 30.

Claughton, P F. 1976. Coal mining in North Devon. *Plymouth Mineral and Mining Club Journal*, Volume 7, No.1.

Claughton, P F. 1993. A list of mines in north Devon and west Somerset.

Cleaver, D, Coulter, J, Slade, P, Waters, E and White, D. 1994. *Bideford Black. The history of a unique local industry*. Sound Archives North Devon.

Cubbon, B. 2008. *Bideford Black*. Pages 159-169 in *The Mine Explorer*, Volume 6 (Cumbria Amenity Trust Mining History Society – CATMHS).

De la Beche, H T. 1834. On the anthracite found near Biddeford [sic] in North Devon. *Proceedings of the Geological Society of London*, 2 (37), pages 106-107.

De la Beche, H T. 1839. *Geology of Devon and Cornwall and west Somerset*. (London: HMSO).

Dines, H G. 1947. Note on Bideford Black. Unpublished typescript report in British Geological Survey files, dated 4 December, 1947.

Edmonds, E A, Williams, B J and Taylor, R T. 1979. Geology of Bideford and Lundy Island. *Memoirs of the Geological Survey of Great Britain. Explanation of sheet 292, with sheets 275, 276, 291 and part of sheet 308*. (London: HMSO). (Page 116).

Edmonds, E A, Whittaker, A and Williams, B J. 1985. Geology of the country around Ilfracombe and Barnstaple. *Memoirs of the British Geological Survey. Explanation of sheets 277 and 293*. (London: HMSO). (Page 78).

Frederick Sherrell Ltd. 1995. A preliminary desk study on mineral workings in the vicinity of the proposed Bideford East-the-Water industrial link road. Report No. 1672.

Frederick Sherrell Ltd. 1996. Bideford East-the-Water proposed housing development. Initial desk study report relating to old mine workings. Report No. 1685.

Gill, Donald. 1941. Report on visit to Bideford Black Pigments Ltd., Bideford Devon. Unpublished typescript report in British Geological Survey files, dated London, 27 October, 1941.

Goaman, M. 1968. *Old Bideford and District*. (Bristol: E M Cox and A G Cox).

Grant, A and Christie, P. 1987. *The book of Bideford*. (Buckingham: Barracuda Books Ltd).

Hall, T M. 1875. Notes on the anthracite beds of North Devon. *Report and Transactions of the Devonshire Association*, Volume 7, pages 367-375.

Langton, H M. 1928. *Mineral black*. Reprinted from an article in *The Industrial Chemist* for Devon Anthracite Limited, 2 Barnstaple Street, Bideford. (6 pages).

Lewis, S. 1811. *A topographical dictionary of England*. (4th Edition). (London: S Lewis & Co).

Lysons, S and Lysons, D. 1822. *Topographical and Historical Account of Devonshire. Magna Britannia*.

Morgans, H M. 1927. Report [on north Devon anthracite and Bideford Black for Devon Anthracite Ltd]. 21 pages plus maps and figures. Dated 4 February, 1927. Unpublished typescript in British Geological Survey files.

Morgans, H M. 1930. A dressing plant for 'Bideford Black'. *Transactions of the Institute of Mining and Metallurgy*, Volume 39, pages 653-662.

Newell Arber, E A. 1905. The fossil flora of the Culm Measures of north-west Devon, and the palaeobotanical evidence with regard to the age of the beds. *Philosophical Transactions of the Royal Society of London*, Volume 197, pages 291-325.

Pigot's *Directory*, 1823 and 1830.

Sedgwick, A and Murchison, R I. 1839. Classification of the older stratified rocks of Devonshire and Cornwall. *London and Edinburgh Philosophical Magazine and Journal of Science*, Volume 14, pages 241-260, and page 354.

Sedgwick, A and Murchison, R I. 1840. On the physical structure of Devonshire, and on the subdivisions and geological relations of its older stratified deposits etc. *Transactions of the Geological Society of London*, Series 2, Volume 5, Part 3, pages 633-703.

Strong, H W. 1889. *Industries of north Devon*. (Barnstaple).

Vowler, F. 2004. Old King Coal. *Atlantic Coast Express Magazine*, Part 1 (No 58), pages 6-7; Part 2 (No 59), pages 8-9; Part 3 (No 60), pages 14-15.

Walcot Gibson. 1920. *Coal in Great Britain*. (London: Edward Arnold). (page 10).

White, W. 1850. *History, gazetteer and directory of Devonshire*.

Chapter 4. The 'Scythestone Hills': The Whetstone Mines of the Blackdown Hills, East Devon

Anon. 1854. *Harper's New Monthly Magazine*, Volume 9, pages 73-75.

Anon. 1994. Devon: home of the best sharpening stone. *Geology Today*, Volume 10, pages 171-172.

Blackmore, R D. 1894. *Perlycross*. (London: Sampson Low). Pages 93-97.

Chalk, E S. 1910. The town, village, manors, parish and church of Kentisbeare. *Report of the Devonshire Association for the Advancement of Science*, Volume 43, pages 278-345.

Chalk, E S. 1910. The manor, parish and churches of Blackborough. *Report of the Devonshire Association for the Advancement of Science*, Volume 43, pages 346-360.

Chalk, E S. 1934. History of Kentisbeare and Blackborough. *Devonshire Association*.

Chalk, R S. 1979. *Blackborough 1904-1937, a pictorial record*. [A scrapbook compiled by the Revd R S Chalk, son of the Revd E S Chalk, rector of Blackborough 1904-1936. Devon Record Office reference 3269Z/1].

De la Beche, H T. 1839. *Geology of Devon and Cornwall and West Somerset*. (London: Longmans). Pages 516-517.

De Luc, J A. 1811 (based on a visit in 1806). *Geological Travels. Travels in England*. Volume 3, pages 24-27. (London: Rivingtons).

Downes, W. 1882. The zones of the Blackdown Beds, and their correlation with those at Haldon, with a list of the fossils. *Quarterly Journal of the Geological Society of London*, Volume 38, pages 75-92.

Dunford, C M (editor and compiler) 2000. *Broadhembury. A picture of our parish*. (Cullompton: Avocet Press). [Whetstones, page 60].

Edwards, R A.1993. Appendix on the geology of the whetstone beds. Pages 106-112 in Stanes, 1993.

Edwards, R A and Gallois, R W. 2004. Geology of the Sidmouth district – a brief explanation of the geological map. *Sheet Explanation of the British Geological Survey. 1:50 000 sheets 326 and 340 Sidmouth (England and Wales)*. (Keyworth: British Geological Survey).

Fitton, W. 1836. On the strata below the Chalk. *Transactions of the Geological Society of London*, 2nd Series, Volume 4, pages 236-238 [Observations in Devon apparently made in 1825].

Fitzpatrick, A P, Butterworth, C A and Grove, J. 1999. Prehistoric and Roman sites in East Devon: the Honiton to Exeter Improvement Scheme 1996-9. Vol. 1. Prehistoric sites in Wessex. Archaeology Report No 16.

Hinde, J G. 1885. Beds of sponge remains in the Lower and Upper Greensand of southern England. *Philosophical Transactions of the Royal Society of London*, Volume 176, pages 403-453.

Hutchinson, P O. 1854. Sketch books. Devon Record Office Z 19/2/8 a-f.

Hutchinson, P O. 1874. Journals. Devon Record Office Z 19/36/16d.

Jukes-Browne, A J and Hill, W. 1900. The Cretaceous rocks of Britain. Vol. I. The Gault and Upper Greensand of England. *Memoir of the Geological Survey of the United Kingdom*.

Kirwan, R. 1866? An account of an excursion of the Exeter Naturalists' Club to the Blackborough whetstone mines. Source uncertain; information obtained from a photocopy in the Broadhembury local history file at Cullompton Library. Possibly from a newspaper report.

Knight, R. 1836. Hone Stones. *Transactions of the Society of Arts*, Volume 50, pages 231-237.

Lysons, S and Lysons, D. 1822. *Topographical and Historical Account of Devonshire. Magna Britannia*. Page 294.

Milles, J. C. 1750. *Milles MS. Parochial Histories*. Bodleian Library, Top Dev.

Moore, D T. 1978. The petrography and archaeology of English honestones. *Journal of Archaeological Science*, Volume 5, pages 61-73.

Moore, T. 1829. *The history of Devonshire*. (London: Jennings). Page 363.

Polwhele, R. 1797. *History of Devonshire*, Volume 1. Pages 65-66.

Rugg, D V. Not dated. *In and around Blackborough*. (Privately printed).

Snell, F J. 1904. *Early Associations of Archbishop Temple*. (London: Hutchinson). Pages 11-21.

Snell, F J. 1911. *The Blackmore Country*. 2nd Edition. (London: Black).

Stanes, R G F. 1993. Devonshire Batts. The Whetstone mining industry and community of Blackborough, in the Blackdown Hills. *Report of the Transactions of the Devonshire Association for the Advancement of Science*, Volume 125, pages 71-106.

Stanes, R G F. 2000. Leases for whetstone pits in Broadhembury. *Devon and Cornwall Notes and Queries*, No. 96, Volume 38, pages 252-253.

Tamplin, R. 2002. *To what extent is the 19th century whetstone mining industry of Blackborough, East Devon, reflected in the local landscape?* A study submitted as coursework for A-level Archaeology, Sidmouth Community College. Unpublished MS, in the author's collection.

Taylor, J D, Cleevely, R J and Morris, N J. 1983. Predatory gastropods and their activities in the Blackdown Greensand (Albian) of England. *Palaeontology*, Volume 26, Pt.3, pages 521-553.

Tucker, D G. 1983. Ayrshire hone stones. *Ayrshire Archaeological and Natural History Society*, Volume 4, pages 5-47.

Ussher, W A E. 1906. Geology of the country between Wellington and Chard. *Memoir of the Geological Survey of Great Britain*. (London: HMSO).

Vancouver, C. 1808. *General View of the Agriculture of the County of Devon*. (London: Phillips). Pages 51 and 72-73.

Woods, M A and Jones, N S. 1996. The sedimentology and biostratigraphy of a temporary exposure of Blackdown Greensand (Lower Cretaceous, Upper Albian) at Blackborough, Devon. *Proceedings of the Ussher Society*, Volume 9, pages 37-40.

Woodward, H B. 1899. Blackdown. *Proceedings of the Geologists' Association*, Volume 16, pages 143-144.

Woolmer, S. 1831. *Gentleman's Magazine*, Volume 24, New Series, Part 1, pages 19-22.

Chapter 5. 'Cathedrals of stone': The Beer Stone Mines of East Devon

Anon. 1911. Reviving an old industry. The Beer Stone Quarries. Interesting Historical Associations. Reprinted from an article in *Pulman's Weekly News*, 21 November 1911. Clinton Devon Estates archive, reference 22170.

Anon. Not dated. *St Michael's Church, Beer. Brief history and guided tour.* [booklet]. (Contains information on the use of Beer Stone and other building materials in the church).

Building Research Establishment. The BRE/British Stone list. http://projects.bre.co.uk/ConDiv/stonelist/beer.html

Davies, G M. 1919. Chromite in the Beer Stone. *Geological Magazine*, Volume 6, pages 506-7.

De la Beche, H T. 1826. On the Chalk and sands beneath it (usually termed Greensand) in the vicinity of Lyme Regis, Dorset, and Beer, Devon. *Transactions of the Geological Society of London*, Series 2, Volume 2, pages 109-118.

Devon County Minerals Local Plan, Adopted Plan: Part B, June 2004. *Beer*, pages 101-105.

Dove, J. 1994. *Exeter in Stone*. Thematic Trails. South Devon Thematic Trail No. 4.

Erskine, A M (editor and translator) 1981. The accounts of the fabric of Exeter Cathedral 1279-1353. *Devon and Cornwall Record Society*, New Series, Volume 24.

Groves, A W. 1931. The unroofing of the Dartmoor Granite and the distribution of its detritus in the sediments of southern England. *Quarterly Journal of the Geological Society of London*. Volume 87, pages 62-96.

Jukes-Browne, A J. 1903. The Cretaceous rocks of Britain. Volume 2. The Lower and Middle Chalk of England. *Memoir of the Geological Survey of the United Kingdom*. (London: HMSO).

Lysons, S and Lysons, D. 1822. *Topographical and Historical Account of Devonshire. Magna Britannia*.

Massey, P E. 1882. Observations on Beer Stone. (Exeter).

Newberry, J. 2002. Inland flint in Prehistoric Devon: sources, tool-making quality and use. *Proceedings of the Devon Archaeological Society*, Volume 60, pages 1-36.

Planel, P. 2008. Beer Quarry Caves Survey Project. Part 2. Inventory of documentary sources. Report for Devon Historic Environment Service. Unpublished TS, 8 pages.

Polwhele, R. 1797. *History of Devonshire*, Volume 1. Page 65.

Scott, J and Gray, G. Not dated. *Out of the darkness*. A brief history and description of the Old Quarry, Beer.

Torrance, J. In press. *On the history of lime-burning in East Devon*. Devon History Society.

Vancouver, C. 1808. *General View of the Agriculture of the County of Devon*. (London: Phillips).

Chapter 6. Ball Clay: The Mines of North and South Devon

Anon. 1964. Ball clay production in south Devon. The operations of Watts, Blake, Bearne & Co. Ltd. *The Quarry Managers' Journal*, August 1964, pages 301-308.

Anon.1976. The quarrying and processing of ball clay. *The Quarry Managers' Journal*, August 1976, pages 43-45.

Ball Clay Heritage Society. http://www.clayheritage.org

Ball Clay Heritage Society. 2003. *The Ball Clays of Devon and Dorset*. (Cornish Hillside Publications).

Brackenbury, C. 1931. Clay mining in South Devon. *Transactions of the Institution of Mining and Metallurgy*, Volume 40, pages 238-277.

Bristow, C M. 1968. The derivation of the Tertiary sediments in the Petrockstowe Basin, north Devon. *Proceedings of the Ussher Society*, Volume 2, pages 29-35.

British Geological Survey. 1975. Ball Clay. *Mineral Dossier No 11*. (Keyworth: British Geological Survey).

British Geological Survey. 2011. *Ball Clay*. Mineral Planning Factsheet. (Keyworth: British Geological Survey).

British Geological Survey. 2010. *United Kingdom Minerals Yearbook 2009*. [Ball Clay pages 24-25]. (Keyworth: British Geological Survey).

Bulley, J A. 1955. The beginnings of the Devonshire ball-clay trade. *Transactions of the Devonshire Association*, Volume 87, pages 191-204.

Chandler, M E J. 1957. The Oligocene flora of the Bovey Tracey lake basin, Devonshire. *Bulletin of the British Museum (Natural History)*, 3, pages 71-123.

Chandler, M E J. 1964. The lower Tertiary floras of southern England, Part IV. *Bulletin of the British Museum (Natural History)*, pages 77-79.

Dunsford, M. 1800. *Miscellaneous observations in the course of tours through several parts of the West Country*.

Edwards, R A. 1970. *The geology of the Bovey Basin*. Unpublished PhD thesis, University of Exeter.

Edwards, R A. 1976. Tertiary sediments and structure of the Bovey Basin, south Devon. *Proceedings of the Geologists' Association*, Volume 87, pages 1-26.

Edwards, R A and Freshney, E C. 1982. The Tertiary sedimentary rocks. Chapter 9 in *The geology of Devon*. Durrance, E M and Laming, D J C (editors). (University of Exeter).

Ewans, M C. 1964. *The Haytor Granite Tramway and Stover Canal*. (Dawlish: David & Charles).

Freshney, E C. 1970. Cyclical sedimentation in the Petrockstowe Basin. *Proceedings of the Ussher Society*, Volume 2, pages 179-189.

Freshney, E C, Beer, K E and Wright, J E. 1979. Geology of the country around Chulmleigh. *Memoir of the Geological Survey of Great Britain*, 1:50 000 geological sheet 309. (Ball clay, pages 56-57).

Heer, O. 1862. On the fossil flora of Bovey Tracey. *Philosophical Transactions of the Royal Society of London*, Volume 152, pages 1039-1086.

Key, J H. 1862. On the Bovey deposit. *Proceedings of the Geological Society of London*, Volume 18, pages 9-20.

Lysons, S and Lysons, D. 1822. *Topographical and Historical Account of Devonshire. Magna Britannia*.

Maton, W G. 1797. Observations relative chiefly to the natural history, picturesque scenery and antiquities of the western counties of England. (Salisbury and London: J Easton). Volume 2, page 106.

Messenger, M J. 1982. *North Devon Clay. The history of an industry and its transport*. (Twelveheads Press).

Messenger, M J. 2007. *North Devon Clay. The story of an industry and its railways*. (Twelveheads Press).

Murray, J W. 1992. Palaeogene and Neogene, In: Cope, J C W, Ingham, J K and Rawson, P F. (editors). *Atlas of Palaeogeography and lithofacies*. Memoir 13, pages 141-147, Geological Society, London.

Polwhele, R. 1797. *History of Devonshire*, Volume 1.

Risdon, Tristram. Published 1714, written 1605-1630. *The chorographical description or survey of the county of Devon*.

Rolt, L T C. 1974. *The Potters' Field. A history of the South Devon ball clay industry*. (Newton Abbot: David & Charles).

Scott, A. 1929. Ball clays. *Memoir of the Geological Survey. Special Reports on the Mineral Resources of Great Britain*. Volume 31. (London: HMSO).

Vancouver, C. 1808. *General View of the Agriculture of the County of Devon*. (London: Phillips).

Vincent, A. 1974. *Sedimentary environments of the Bovey Basin*. Unpublished M Phil thesis, University of Surrey.

Wilkinson, G C and Boulter, M C. 1980. Oligocene pollen and spores from the western part of the British Isles. *Palaeontographica*, B175, 27-83.

Chapter 7. Bovey Coal: The Lignite Mines of Bovey Tracey

Adams, B. 2005. *Bovey Tracey Potteries, Guide and Marks*. (Bovey Tracey: House of Marbles).

Adams, Amery W. 1946. The old Heytor granite railway. *Transactions of the Devonshire Association*, Volume 78, pages 153-160.

British Geological Survey. Various files and letters relating to lignite and montan wax in the Bovey Basin, formerly held at the office in Exeter.

Cawley, C M and King, J G. 1946. The extraction of ester waxes from British lignite and peat. *Department of Scientific and Industrial Research. Fuel Research Technical Paper* No 52. (London: HMSO).

Davies, W Brynmor. 1947. Report to the British Lignite Products Ltd on the lignite resources and plan of mining developments proposed in the Bovey Tracey lignite basin, south Devon.

De Luc, J A. 1811 (based on a visit in 1806). *Geological Travels. Travels in England*. (London: Rivingtons).

Devon Record Office. 1508M/Devon/Mining/25. State of Accounts of the Coal Works and Lime Trade at Bovey Tracey. (1756-1764).

Devon Record Office. 1508M/Special subjects: Mining 25. Courtenay family accounts, 1775.

Ewans, M C. 1964. *The Haytor Granite Tramway and Stover Canal*. (Dawlish: David & Charles).

Fiennes, Celia. Manuscript about 1700, published 1888. *Through England on a Side Saddle in the time of William and Mary*. Also more easily available as: Morris, C. (editor) 1947. *The Journeys of Celia Fiennes*. (London: Cresset Press); or Morris, C. (editor) 1982. *The Illustrated Journeys of Celia Fiennes c1682-c1712*. (Exeter: Webb & Bower).

Fox, J, 1948. Lignite. Notes on the characteristics of Devonshire lignite with particular reference to steam raising. *Claycraft*, Volume 22, pages 125-130.

Heer, O. 1862. On the fossil flora of Bovey Tracey. *Philosophical Transactions of the Royal Society of London*, Volume 152, pages 1039-1086.

Hatchett, C. 1797. Observations on bituminous substances, with a description of the varieties of the elastic bitumen. *Transactions of the Linnaean* Society, Volume 4, page 129.

Hatchett, C. 1804. Observations on the change of some of the proximate principles of vegetables into bitumen; with analytical experiments on a peculiar substance which is found with the Bovey Coal. *Philosophical Transactions of the Royal Society of London*, Part 1, page 396.

Hoskins, W G. 1972. *Devon*. (Newton Abbot: David & Charles).

Kelly's Directory. 1893 (pages 11 and 70; 1919 (page 22).

Kennedy, V (compiler). 2004. *The Bovey Book. The story of a Devonshire town in words and pictures*. (Bovey Heritage Trust).

Manchester Guardian. 12 November 1947. Mining at Bovey Tracey. New Exploitation of Britain's Main Lignite Deposit.

Maton, W G. 1797. *Observations relative chiefly to the natural history, picturesque scenery and antiquities of the western counties of England.* (Salisbury and London: J Easton). Volume 2.

Milles, J. 1760. Remarks on the Bovey coal. *Philosophical Transactions of the Royal Society of London*, Volume 51, page 534.

Moore, T. 1829. *The history of Devonshire from the earliest period to the present.* (London: Robert Jennings). (Pages 266-280).

Parish, C W. 1947. *The Creation of an Industry.* [booklet].

Parkinson, J. 1804. Volume I. Letter XI, page 104; Letter XII, pages 126-129. London.

Pengelly W. 1862. The lignites and clays of Bovey Tracey. *Philosophical Transactions of the Royal Society of London*, Volume 152, pages 1019-1038.

Pengelly, W and Heer, O. 1863. On the lignite formation of Bovey Tracey, Devonshire. (Reprinted from the *Philosophical Transactions*, Part II, 1862). (London: Taylor and Francis).

Polwhele, R. 1797. *History of Devonshire*, Volume 1. Page 65.

Scammel, R. 1804. Particulars respecting Bovey Coal, dug at Bovey, near Chudleigh, Devonshire. Letter XII, pages 126-129 in Parkinson, 1804. *Organic remains of a former world.*

Scott, P H W. 1979. *The geology and mining of the lignite in the Bovey Basin, Devonshire.* Camborne School of Mines.

Strahan, A. 1918. Special Reports on the Mineral Resources of Great Britain. Vol. VII: Lignites, Jets, Kimmeridge Oil-shale, Mineral Oil, Cannel Coals, Natural Gas. Part I: England and Wales. *Memoir of the Geological Survey of Great Britain.*

Tregoning, L. 1983 (new edition 1993). *Bovey Tracey. An ancient town. Its story and legend.* (Bovey Tracey: Cottage Publishing).

Vincent, A. 1974. *Sedimentary environments of the Bovey Basin.* Unpublished MPhil thesis, University of Surrey.

Weddell, P J and Westcott, K. 1986. The Bovey Tracey Pottery kilns. *Devon Archaeological Society, Proceedings*, No. 44, pages 143-162.

White, W. 1850. *History, Gazetteer and Directory of Devonshire.* Sheffield (page 469).

Wynne, J N. 1947. Lignite mining in Devon. *Mining Magazine*, Volume 77, pages 329-336.